Move to Reality:
A Statistics Teacher's Guide
to
The Intersection of Science and Faith

by
Bill Richardson

ABOUT THE AUTHOR

The author is a statistician and high school teacher of 30 years. He possesses a B.S. in Mathematics and a M.S. in Applied Statistics and is a multiple *Teacher of the Year* recipient. By 1986, his life had been profoundly and negatively impacted by misguided Christian teaching and decisions that arose from it. He walked away from church-life as a result. Many things bothered him, including the treatment of certain people groups and how to reconcile the extent of pain and suffering in this life if a loving God is truly in control. Like many, he simply could not accept a spiritual perspective that does not line up with reality.

Tragedy in lives of others eventually led him to question and seek again. In 1990 he began to investigate whether one could take a logical ascent to establishing the existence of a God, the existence of an afterlife, and the reason for so much pain and suffering. He soon recognized that the bias he and others have acquired over a lifetime makes it difficult to question, correct, and substantiate critical spiritual components. He experienced the cognitive dissonance that accompanies such a journey. For decades he has been involved in various efforts to improve the lives of others educationally, financially, emotionally and spiritually. But with his ability to connect spiritual dots while being accountable to universally accepted science and history, it is introducing others to the God of the universe that most compels him.

INTRODUCTION

For 30+ years the author has worked on developing an ideology – a spiritual ideology – that is accountable to accepted Scientific discovery and to the Scriptures – the Jewish Tanakh, as identified by the Apostle Paul. The Tanakh, basically the Old Testament of The Bible, is accepted in whole or in part by the Abrahamic faiths, including Christianity, Judaism, and Islam.

When Science confirms with many factors the universe is 13.7 billion years old but Scripture seems to suggest otherwise, a more careful look at Scripture must occur. Finding a Christian astrophysicist like Dr. Hugh Ross helps to solve the conundrum. He explains why the original Hebrew in the Book of Genesis describes long periods of time (yom) and not 24-hour days. Exploring this possibility confirms harmony in Science and Scripture and can lead scientists like Ross to faith.

When Scripture yields many fulfilled prophecies of the Messiah but some in the Scientific community assume these were written after the life of Christ, a careful look at History is necessary. Each of the books in the Tanakh were written before 400 BC. Therefore, the prophecies cannot be dismissed and are important factors of reality.

Better grasping our human nature from the writings of Nobel Prize Laureate Daniel Kahneman reveals how easily group-think and confirmation bias arise in Scientific, Religious, and Political mindsets. It helps us understand how to wrestle with our ideologies and tendencies. And it helps clarify why error in key Scientific assumptions and error in some modern Christian doctrine can so easily occur.

When contradictions exist within and between Science and Scripture and History, one can continue to investigate and make correct adjustments to produce an ever-improving and substantiated Spiritual ideology. Scientific information acquired can produce a series of epiphanies and fascination as one discovers how extraordinary our planet and solar system are and the various moments in the past four-billion years when Earth nearly became an ice ball or narrowly avoided the end of life. The specialized research of thousands of astrophysicists and chemists in recent decades reveals hundreds of characteristics required for a planet to spawn life and evolve with protection over billions of years. Assumptions still shared in many classrooms are no longer supported.

Connections in Scripture – some rarely realized – can amaze and ignite passion as a faith-producing precision is discovered. The God most of us desire is confirmed – a God of all people groups, and a God of grace and patience. And a more accurate perspective on pain, suffering, and the existence of evil is identified, including why they are necessary.

The result of this book is a proof of God's existence. It is a precise perspective on Science and Faith. It helps us appreciate our purpose in life and the joy and peace we can experience despite the pain and brutality that life and others may inflict. And it helps us better appreciate the beautiful souls we turn to in times of question or crisis and the likelihood that they experienced trauma or grief in the formation of their wisdom.

DEDICATION

*Dedicated to the millions lost or harmed
due to our ignorance*

Table of Contents

Chapter 1

Why a Statistics Teacher

I began teaching in 1988. In 1994 I earned a M.S. in Applied Statistics. We sold our house and prepared to move to Chicago or New York or Atlanta, where many high-paying jobs existed for Statisticians. As I weighed options, I kept having my heart tugged in the direction of working with students and making a difference in their lives. For better and worse, I decided to remain a teacher while trying to make extra money on the side.

As I wrestled with my beliefs and continued to search for a spiritual ideology built upon accountability, I was fortunate to teach in schools with diverse populations by race, economics, politics, and religion. As a statistician and one with a high degree of empathy, I gained a wide range of impactful information and was better able to understand and connect with students of nearly every demographic.

At one point I started a group for black males at my high school. We had no black male teachers in the core subjects despite having nearly 300 black students in the school population. I began with a core of black student leaders from grades 10 – 12. Once we developed a vision, we went in pairs to personally invite others.

I will never forget the moment when we invited Christian to join our group. He was in low-level classes and seemed to move a bit slow, but I soon learned he was booted from a sister school for gaining access to the computer system and altering attendance information for students. Well, by golly, I felt I had to utilize his technical expertise. He became my right-hand man with significant access to the computer system. He eventually paid most of his way through college with IT jobs and recently earned his master's degree from the University of Denver. He is involved in politics and is still active in leading the group we began.

Eighty students and I met weekly to motivate, strategize, and grow. I quickly recruited adult black males from the community to get involved. I am white and was not interested in exploiting the situation. Student backgrounds ranged from former gang members to some of the top students in the school. Everyone had something to contribute. Some questioned why it was necessary to have a separate group for black males. But it was important. Many mindsets, factors, and systems affect blacks in negative and harmful ways. Important discussions took place and guys grew emotionally and relationally. Successful students helped others improve, and GPA's rose significantly. We were like family, and the guys were rock stars at school. The student body felt the energy and had the respect to listen, observe, interact, and learn. It was a healthy school environment. We soon spread to middle schools and elementary schools. The outgoing black President of the NEA heard what was happening and paid us a visit.

In my Statistics classes, students enjoyed hearing my insights into game theory and strategy. I was not a typical teacher. They knew I wanted to help them consider the power of problem-solving and strategizing. I encouraged an entrepreneurial and innovative spirit for those possessing certain traits. The examples and course provided them tools with which to analyze.

In 1994 I began sharing the analysis I did on why NFL teams should be going for touchdowns versus field goals on 4th down inside the five yard-line with at least six minutes to go in a half. Teams willing to do this would average two additional points per attempt. Unlike today's opportunity to utilize the internet, I had to go to libraries and look through microfiche of newspapers to gather the data I needed. Ten years later students came to class on a Monday excited to tell me a professor of Economics at Berkley was interviewed on an NFL pregame show saying the same thing I shared in class. More than twenty years later, FiveThirtyEight and ESPN Magazine touted fourth-down expected value strategy as the single most beneficial strategy affecting the outcome of a professional football game.

In the late nineties, I began sharing my research, strategies, and accomplishments in blackjack. I developed a multi-deck card counting system in 1988 far superior to published systems. My error per card in representing value was 3% compared to other systems at 8% to 16% error. I also shared how I wrote thousands of lines of computer code to simulate play and to optimize betting progressions designed to further limit variance. The fluctuation of money gained and lost in blackjack is far greater than most would think and far greater than blackjack books tend to admit. Due to trimming variance {standard deviation, actually} to half of what successful counters could, I averaged twice the revenue of other card-counting methods using the same size bankroll and maintaining the same level of risk.

I developed a very successful business model making one short blackjack trip per month. As intriguing as it was to analyze the game of blackjack effectively, the analysis leading to the information in this book is far more exciting and important. And it should be noted that blackjack is no longer a viable money-making venture due to factors such as patented continuous shuffling machines used by casinos.

As a teacher with a high level of empathy, I was concerned with the emotional health of my students in addition to their educational and cognitive development. Sadly, during my career, at least 11 of my students died young and many more experienced severe crisis. Three committed suicide. Four died of gunshot wounds. Two died of disease. Two died of alcohol or drug related causes. The information in this book is more the result of student pain and my pain than anything else.

When students feel safe approaching a teacher about problems and when a teacher is sensitive to a student impacted emotionally by other forces, a bond of hope can be established. It is best to listen and not feel a need to have answers. One must avoid statements like, "I understand," or "Everything happens for a reason." There is tremendous potential when we listen and learn and care.

I had the honor of teaching Mike, an incredibly good-hearted student who struggled with shame and inadequacy. He wandered into my office one day after I led an important discussion in class about meaningful issues. He felt safe in candidly sharing aspects of his life. He was suicidal at times. Yes, I let his counselor know of this, and his parents were already aware. Rarely have I heard such sincerity. He asked me spiritual questions, and I replied as best I could at the time. He sensed there was significance in what I shared, but he was too honest to claim acceptance of ideas he could not yet reconcile with reality.

Mike stopped by a few times as a young adult after graduation. Once was to take me for a ride in his Toyota Supra that he had modified. Thankfully, the four-lane street he took me on was empty as he went from zero to 100 miles per hour in *oh my gosh*. Later, he contacted me from Iraq, where he served in the military. When he returned, he stopped by to tell me about this amazing woman he had met. He smiled as he said, "And she's a Christian." Mike married her. He was still seeking for the information about God and a possible after life.

Not too long after that, his Mom contacted me. Mike had committed suicide, and the family wanted me to speak at his funeral. At the time and on many days since, I cried like a baby. Mike was a beautiful human being, and he was gone. When I arrived for the funeral, I asked his wife a very important question. "Did he find the information he needed? Did he accept what Jesus could do for him?" She smiled, and we laughed and cried as she told me about the past six months. Mike was very intelligent. Almost weekly, he would stop the pastor at their church to ask a baffling question. Like me, the pastor loved Mike and his level of integrity. Mike could not fake it. He had to have an intellectual, emotional, and spiritual understanding of Christ's purpose and of our position in the universe with God.

   A few months before he took his own life, Mike accepted what Jesus did for us and committed his heart to him. Anyone with honesty and integrity like Mike will understand this life to be beautiful and brutal. PTSD and life-long issues led Mike to the desperate point of taking his own life. His death made it even more clear of my need to continue challenging my understanding of things and to produce a spiritual ideology with greater and greater levels of accountability to scripture, science, and history. There was much evidence and clarity already attained, but I needed more. And I needed to find ways to better spark the interest of students to have hope and to ponder our existence here. To do this, I had to devote time and study while continuing to have the discipline to question my beliefs. And I had to operate with the knowledge of what a teacher can and cannot share in a public classroom.

   The respect I have for all people groups made it easy for my students and me to have meaningful class discussions on vital issues. After 9/11, I did something that was a bit risky. At that time and in addition to the terrorist attacks, there was an individual mailing anthrax to certain government officials. I walked into class shortly after 9/11 and said I had mailed away for an incredible math game. "It arrived yesterday in the mail." I showed them the manila envelope and began opening it. As I did, a white powder fell to the ground. The seniors froze in their seats for a moment. But I quickly stated. "All right, you clowns! It's powdered sugar! Who's going to send anthrax to some goofy math teacher in suburbia?"

I proceeded to share the notion of fears that are rational and fears that are irrational. I explained that the planes flying into the Twin Towers and the Pentagon were part of a diabolical plan that took months to pull off. "Will we have a terrorist attack on our soil again?" I asked. "Probably. But it will likely be several years not several months or weeks before something this dramatic occurs again."

During that time, the President asked Americans to continue flying. But I shared my concern that too many Americans would not fly due to irrational fear. "If too many are afraid to fly, the terrorists will achieve an even greater goal – a significant impact on the economy." And in fact, in the weeks following the attack, tourism and travel were heavily impacted.

Additionally, I mentioned a conversation I started with a gentleman of Middle Eastern decent at a health club. I knew many were suspicious of anyone who looked like him. "How are you holding up with all of this?" I asked. "Are you getting suspicious looks from people?" He replied, "You have no idea. I was at the gas station filling my tank, and I could feel people glaring at me. I looked around and, sure enough, they were." We talked for a bit. Mostly I just listened and then expressed my sorrow that so much distrust of fellow Americans existed.

I brought a large card to class for any interested student to sign. I read the letter I was enclosing and told the students it would be sent to the local Islamic Center. The essence of the letter was to let members of their community know that not everyone had lost their minds. It emphasized that we are all brothers and sisters in the United States and that we were saddened by the treatment and mistrust toward many in their community. Over the next few years, I continued to mention all of

6

this to students and to discuss how we can use our rational thought and objective evidence to counter irrational fears and avoid prejudice.

A few years into sharing this, one of my Statistics students, Jo Jo, blurted out, "That was you? That made such a difference when we received that card and letter!" Jo Jo is a bright person who strives to make a positive impact on the world; and she was delighted to make the connection. It was another powerful, spontaneous moment that brought the class together and likely eroded bias and prejudice for some students.

I hesitate to share too many painful details of my life due to the negative portrayal it would have on a few individuals. I grew up in a home like many, where the standards were set too high. Shame and insecurity played a major role throughout the years. An evangelical Lutheran family had an impact on me during my teenage years as did my older brother, who joined the Pentecostal movement. I chose a college where I could play baseball and attend a large Pentecostal church.

The pastors at the church were teaching the end of the world and that we were about to enter the tribulation. I was swept up by the emotion and intrigue of it all. As the school year ended, I felt college and baseball should be abandoned in exchange for a more evangelical life. I moved to another state and began to attend a very small church with young leadership. This church was into *"exposing the forces of darkness."* I was going further down the rabbit hole of paranoid ideas that lacked accountability to Biblical teaching, science, history, and reality in general. I alienated many of my long-time friends by condemning their ideologies and sharing my newly acquired, fanatical beliefs.

During this time at the church, many young people were getting married. In one evening, someone I considered a mentor talked me into proposing to a girl I hardly knew. A few days later I realized it was a huge mistake, but adults in my life went silent or guilted me into going through with it. My fanatical, inaccurate devotion to God left me so focused on treating others well that I married so I would not hurt this person. The marriage was a disaster that lasted 15 years but with ongoing, significant consequences more than twenty years after the divorce.

Two years into the marriage my wife and I realized church and life were too emotional. We ceased attending church and went back to school. I began teaching middle school in 1988 and moved to the high school level the following year.

Something significant happened in 1990. A colleague and his wife lost their little girl to meningitis despite hundreds praying for her. I recall ranting through the ceiling at a God I was not sure existed. "Why? Why allow an 18-month-old child to die?" I told him I was putting him on trial. His existence would have to line up with reality. I merely wanted the truth. I began a journey that has now lasted more than thirty years. An ideology built primarily on emotion and running with the religious crowd was unhealthy. Either God and a plan exist, and I could renew my mind to a harmony with science, observable reality, and following God; or I could move beyond the investment I put into religion and experience a full life of relationships, service, joy, and growth despite understanding no God or plan existed.

My oldest boy was born in 1991, and my youngest followed the next year. I spent 1993 and 1994 earning my master's degree while teaching. By the end of the nineties, there was no way my wife and I could continue our marriage together.

Like many who experience divorce, I felt crushed. I remember sitting in church not wanting to be noticed as I buried my head in my hands and tried to control the tears that would not stop coming. It was at that time I met Pastor Bruce Ramsey, who instilled the appreciation for the pain and hurt we experience as we journey through life. He conveyed the understanding of the benefit of embracing whatever the past, present, and future hold. And he shared a statement I have often relayed: *"Life is a struggle. Struggle well."*

My boys were six and eight years old when the divorce occurred. I never dreamed I would be a divorcee and expose my children to the pain and complexity of it. I eventually fell in love and have been married to my current wife for twenty years. My first wife died of cancer years after the divorce. The issues surrounding the divorce, her battle with cancer, and her eventual death took a tremendous toll on all of us. Depression and anxiety would be a part of life for years to come.

My book would never have been written without the pain that kept me learning and refining. It required listening to the wisdom and questions of my children and my students; and it required the sincere desire to aid those going through painful circumstances. For thirty-one years as a teacher, I was in the unique position of tuning in to a beautiful diversity of teenage students – what they knew and didn't know, what they thought they knew, their fears, their questions, their hopes, and their assumptions. And I was able to interact with exceptional teachers – some of whom were grossly over-qualified and powerful classroom teachers and role models.

In the chapters that follow, there will be deconstruction of ideology by purging unaccountable ideas, followed by a reconstruction accountable to accepted scientific discovery, scripture, and history. Human nature must be recognized for its flaws and how it affects us individually and corporately. Group-think in the scientific and academic communities must be addressed as do errant belief in the religious sector. I will mesh sound science with a faithful accountability to Scripture. The result is a faith-affirming clarity.

Chapter 2

The Trouble with Human Nature

I taught AP Statistics and Algebra 1 for much of my 31-year teaching career. Though personal finance was not part of the AP Statistics curriculum, I chose to create a meaningful exercise involving a monthly budget. This was the situation:

You are 23 years-old and single. You just graduated college with a Business degree. You live in a nice, two-bedroom apartment in a middle-class suburb, because you want an office in addition to a bedroom. You bought a new car. On the blank monthly budget sheet provided, estimate as best you can the cost of these items along with food, phone, utilities, insurance, internet, clothes, giving, and other typical expenses. Add the expenses to get the total of your monthly budget.

They were not allowed to consult anyone or use smart phones to look up costs... yet. After 10 minutes I asked for some quick estimates from students regarding the cost of the apartment and the monthly food costs. I did this for two reasons. The first was that, inevitably, some students would have absolutely no clue about money and grossly *over-estimate* the cost of food or an apartment, which was hilarious. And the second was that, inevitably, some students would have absolutely no clue about money and grossly *under-estimate* the cost of food or the apartment, which was equally hilarious.

They spent time in groups at the end of class sharing and improving estimates. Their homework was to consult with parents or others and to research things online such as the typical cost of a two-bedroom apartment and monthly payments on a new car with an $18,000 loan balance.

I greeted them the next day with a spreadsheet projected on the front board. After they consulted in groups about the various costs, we as a class completed the spreadsheet. Students were then informed of the median entry salary for someone with a Business degree. Taxes were deducted, and the students now realized we were over budget by more than 50%. Back into groups to discuss sacrifices to trim the budget – used cars, roommates, cheap one-bedroom apartments in scary neighborhoods, and lots of Ramen noodles. Finally, a balanced budget.

Just as they thought we were finished, I slapped my forehead and said, "Oh no! I forgot something. You decided to go all four years to an expensive public college and amassed $60,000 in student loan debt. Therefore, you will pay $750 per month for the next **ten years!** The result? You're moving back in with your parents or living in a van down by the river."

We finished with a very serious discussion of researching job markets on the internet – particularly utilizing the U.S. Bureau of Labor Statistics website. What are the earnings for various degrees? What jobs are projected to grow in demand? If your desire is to pursue a career which does not typically pay well, you may need to spend the first year at a community college to save money. Some students may achieve success and fulfillment by pursuing careers not requiring college. It was easy to justify a little time in class for this.

As a Statistician teaching in suburban high schools, it was surprisingly difficult to get district and school administrators to question the status quo despite providing them with rich, objective evidence suggesting the ineffectiveness and dangers of certain practices and mindsets. It is not that they lacked intelligence. It was more a lack of analysis training and of human susceptibility. Inaccurate assumptions often tainted their decision-making process. The consequences for students were numerous.

Schools are often evaluated by test scores, number of AP Exams taken, and percentage of students advancing to college. While generally good standards to target, these can be detrimental. I often witnessed unrealistic demands placed on students and teachers in attempts to achieve these. Volume of homework often dwarfed efficiency. It was tempting to emphasize rote memorization over higher levels of thinking.

Stress levels ran high at our school. We had two suicides and over 200 suicide-assessed students my final year of teaching. A few measures were put in place for the following year. However, they were built on confirmation bias and a disregard for data suggesting how ineffective such measures have been historically. And the oppressive culture of the demands placed on students and teachers remained in place. Students were encouraged to stretch themselves and take advanced classes in all subjects. Homework loads – particularly for less qualified students – were unhealthy. Too many bought in to the notion that students needed to fill their time with advanced classes, extra-curriculars, four years of foreign language, and high ACT or SAT scores, because it is what colleges require.

Despite test scores well above average in general, students scored relatively low in problem-solving and reading. And the level to which students were dependent upon adults for assistance and decision-making was often observed. The culture ran on the assumption that ultimately it was in the best interest of the students to be pushed hard and to set high expectations. Data and evidence were routinely ignored. Status quo bias and confirmation bias were commonplace.

At one point, I surveyed my students on what they would like adults to know about their situation and stress. Many students were in the midst of family crisis and/or worked jobs outside of school by necessity. Some suffered from physical or mental health issues and expressed how teachers often assumed they were lazy or lacked responsibility. Many had full weekends with church or club activities, sports, or marching band. Yet, they felt several teachers assumed they had many free hours on weekends to complete larger projects or long homework assignments. The stress felt by many was dangerous.

I was able to share this student information with a group of faculty assigned to a professional development session I facilitated. In the session, I asked the faculty to share what they wanted students to know about their situation and stress. Their first concerns were for students, and there was plenty of worry to go around. This is the heart of most faculty. They love their students, worry about them, and want to make a difference in their lives. I had to ask again to please share their concerns for themselves and not students.

Counselors were not sure they could make it to retirement. Their responsibilities have grown and can hardly be met. And despite lacking professional training, they were now asked out of necessity to play the role of mental health workers due to the massive need for it among the student body. One performing arts teacher spoke of getting home late most evenings and having to decide with the little time left: "Should I be a good parent, a good spouse, or a good teacher tonight?" I brought these comments back to class to share with students.

I wish I could say this and other information and data made a difference and that effective changes were made to alleviate these concerns. Obviously, the people in positions of power did not want students to be harmed. They got into education to make a difference. But as is true with most adults, they had little grasp of the power of their human nature to interfere with quality decision making. Beliefs can be deeply held. When it comes to changing a culture or a system, the vital information for positive change is usually available and presented by some one or some group. It is often not the lack of information and viable solutions that prevent progress. Instead, it is the various ways our human nature can betray us individually and corporately. This became more and more apparent later in my teaching career after fighting many important battles armed with valid data and supported solutions.

Confronting our spiritual beliefs and ideologies can also be severely impeded by how we are wired. Therefore, it is important to consider the impact of human nature.

In our brains the prefrontal cortex is engaged when carefully analyzing. However, the limbic region is where we tend to operate. It relies heavily on intuition, our immediate perception of things, and the meaning we have previously constructed to help make sense of what we are encountering.

Daniel Kahneman won the Nobel Prize in Economics for his work on Economic decision-making. He displayed how expert financial analysts make irrational investing decisions despite being convinced the decisions are logical and wise. In his best-selling book Thinking, Fast and Slow, Kahneman compiled well-supported research to paint an accurate picture of how humans think and act. He refers to System 1, the limbic processor, and System 2, the prefrontal-cortex processor. System 1 is predominant. It is intuitive – instinctive. It attempts to quickly assess a situation via intuition, immediate perception, and previously constructed meaning. System 2 is effortful and deliberate in attempting to make a more responsible and accurate assessment of things. However, unless there are certain external stimuli to engage System 2, it is System 1 that is in control and must give permission for System 2 to operate.[1] System 1 is very self-confident and can influence System 2.

Much as in the example of financial experts, this human flaw of being self-assured impedes everyone regardless of intelligence and level of education. All humans are susceptible to reaching inefficient and inaccurate conclusions. Awareness of this trait can help us better assess other ideas and theories. Though belonging to a larger group of like-minded people can help persons feel safe, it can also embolden shared, errant beliefs. Confirmation bias can impede an earnest search for reality. Corporations, news sources, and other groups or individuals can exploit this need-to-belong and thus influence beliefs. Many naturally assume it's the *other guy* that is being duped. Our tendency to *think it's the other guy being duped* and to embrace our intuition is perhaps the best example of our human nature at its core.

How human brains came to be hard-wired this way is not certain. Naturalists would suggest an evolutionary explanation. Over millions of years, species had to operate intuitively and instinctively to protect against immediate danger from predators and nature itself. This would explain why this quick-thinking, intuitive processing is primary in us. We quickly assign meaning to the things we observe. The meaning comes from our past experiences and our own bias, and we are very convinced of our perspective.

Alternatively, many with a religious ideology would claim that human nature is the *sin-nature* God created in us. The idea might be that God created humans with this flawed nature. By placing them in this beautiful yet brutal world, there would be opportunity to find purpose and answers for those willing to seek with integrity and humility. Perhaps our condition is an orchestrated result of both explanations and more. Regardless of how we arrived at our current state of existence, there is much agreement within the scientific community on how the two regions of our brain interact and process.

Most people have been primed to accept information presented if it is shared by an "expert." In many classrooms of the past and present, a teacher is in front of a captive audience telling them what to think – providing the *correct* information. From a young age, people are programmed to trust the experts in front of them.

In History class, for example, instructors and documentaries can look back in time with seemingly perfect vision and share just the right factors and events that led a society into war or financial crisis. Students may emerge with notions of the signs to look for in any situation that will point to economic demise or war. But life is far more complex with wide arrays of factors continually interacting. Looking back on history provides much more clarity than making history. If recognizing warning signs was so easy in real time, our history books would not be filled with so much destruction and turmoil. And **always**, the greatest contributor to history is human nature.

There are many examples of negative *group-think* via human nature. In the 1960's and 1970's, some argued that it was safer <u>not</u> to wear a seatbelt in a car. There was anecdotal evidence of a person thrown from a car and surviving when he would have died if wearing a seat belt. It took years and government mandates for sound, scientific and statistical wisdom to win the debate.

Social media and the internet tend to make human error, confirmation bias, and thus *group-think* more prolific today than ever. Many love conspiracy theories. In the 1960's a writer for Playboy Magazine teamed with an author of a parody religion to create fictitious letters from readers for and against the existence of a secret society called the Illuminati.[2] The authors knew the actual Illuminati began in 1776 and died out soon after. But their hoax gained traction, and today millions of people believe this defunct, fictitious secret society lurks behind the scenes.

Currently, the misinformation surrounding climate change and the role of mankind in it are riddled with the influence and consequence of human nature. For those who have objectively considered scientific discovery around this issue, it is extremely frustrating to observe millions of people who deny it or fail to admit man's role or responsibility to deal with it. The results of climate change will continue to worsen unless a critical mass of governments, corporations, and people can work to reverse trends and prevent some of the threatening consequences.

Recent events surrounding COVID-19 and racial issues have underscored the dramatic ways in which human nature pits persons and groups against one another. People can be easily persuaded by emotional misinformation from sources they trust.

Years ago I began sharing a powerful perspective I developed on life. Though simple, it has been life-changing for my students and me. It is the practice of seeing everyone and most ideas as both *good **and** bad* as opposed to intuitively *good **or** bad*. People tend to categorize others as heroes or villains. This is dangerous. It is more powerful to look both for good and for bad – effectiveness and ineffectiveness – in those we admire and in those we dislike, and in ideas with which we agree or disagree. It creates greater objectivity and an ability to more accurately define problems and develop solutions. We can be fairly certain each of us and every idea are inaccurate to some degree.

Individually, it is powerful to view one's self as both *good **and** bad*. It removes much of the unnecessary pressure to always do good and be good. All of us have traits we despise. And all of us have some inaccurate perceptions of things. When we can search for and identify these, it is not an indication that we are wrong, or that we are failures, or that we are bad. It simply implies opportunity to add more good to our goodness and delete some bad from our badness.

Chapter 3

Six Errant "Christian" Beliefs

At age 20 my zealous roommate and I visited the swimming pool area at our apartment complex. We were passing out leaflets and telling people about Jesus. There was a person a little older than us sitting by himself in the grass outside the fence enclosing the pool area. I walked through the fence gate and over to where he was. He looked very sullen and noticeably depressed. As I approached, he began to talk to me in a quiet and sincere way. "I hear what you guys are sharing over there. I was you three years ago." He slowly lifted a beer to his mouth, took a swig and said, "Be careful, man. It messed up my life." I will never forget the incident and the genuine care in his voice. Deep down I knew there were things I was accepting that deviated from reality. I hoped I would not lose faith like him.

Three years later, I too was at a point of severe depression. The choices made during an overzealous, reckless time would haunt me for the rest of my life.

What follows are six errant beliefs held by millions of Christians today. These beliefs have tremendous potential to harm and to prevent people of integrity from recognizing the brilliance of the purpose of this life and the hope of a life after it. The chapter following this one will consider errant beliefs held by many who consider themselves *Scientific* and the tremendous potential of those beliefs to harm and to prevent people of integrity from recognizing the brilliance of the purpose of this life and the hope of a life after.

One historical figure – perhaps more than any other – provides the greatest insight into an accurate spiritual perspective. Saul of Tarsus was a Roman citizen and a highly educated Jewish Pharisee. [New International Version, Phil 3: 4 – 6] Saul despised Jesus and his followers – considering them to be heretics. He rounded up early Christians to be imprisoned or put to death. But then something remarkable happened. Through various events and ultimately through his knowledge of Jewish scripture, Saul clearly recognized Jesus as the Jewish Messiah. Saul converted, changed his name to Paul, and became the greatest instructor of the Christian faith.

Early Christians feared this man who had destroyed lives of many followers. But over time it became obvious that his conversion was genuine and profound. The Christian New Testament contains many of Paul's instructional letters. He gave up his wealth and stature in order to travel much of the known world and explain why Jesus was the promised Messiah. He was dangerous in the eyes of powerful Jewish leaders, who rejected Jesus as the Messiah. Thus, Paul was a wanted man – escaping death or capture on various occasions. After many years of teaching, church-planting, and missionary work, he was captured and ultimately beheaded for his faith. Of additional significance is the fact that none of the twelve disciples of Jesus recanted. Eleven of them were forced to recant or be put to death. None of them recanted. They were all put to death and in gruesome ways. These men were convinced Jesus rose from the dead.

Paul's instruction is extremely relevant today. There are foundational ideas many churches ignore. One example is the common interpretation of 2nd Timothy 3: 16. It is part of a letter from Paul to a young Timothy, who was a significant figure in the early church. The verse states that, *"All scripture is God breathed."* Many Christians claim this pertains to the entire Bible – Old and New Testament. However, Paul clearly identifies what is meant when he references *all scripture*. Here were his words two verses earlier.

"But as for you [young Timothy], continue in what you have learned and have become convinced of, because you know those from whom you learned it, and how from infancy you have known the holy Scriptures, which are able to make you wise for salvation through faith in Christ Jesus. All scripture is God breathed and is useful for teaching, rebuking, correcting and training in righteousness, so that the man of God may be thoroughly equipped for every good work." (2Tim 3: 14 – 17)

Here Paul defined what the Scriptures were. They do not include the New Testament of the Bible, since it had not been written when Timothy was growing up. The scripture Paul references is the Jewish Tanakh (or Tanach). It is the Jewish acronym for the Torah (Pentateuch, which includes the Law), the Navi (Prophets), and the Ketuvim (Writings, including the Psalms). Luke 24: 44 records Jesus referencing the Tanakh when he stated, *"Everything must be fulfilled that is written about me in the Law of Moses, the Prophets and the Psalms."*

It is quite significant that the scriptures be correctly identified as the Tanakh, because all the Abrahamic faiths (Judaism, Islam, and Christianity) assign special meaning and inspiration to the Tanakh - though Islam does not consider the entire Tanakh to contain truth. Delving deeper into Paul's teachings, he claims that the death and resurrection of Jesus paves the way to eternity for all people groups - not just Jews and professing Christians. (Rom. 2. 12 – 16)

The New Testament must be looked upon differently than the Tanakh. In the four gospels, it contains various perspectives of the life of Jesus. The Book of Acts is a record of the early church. Paul's letter to the Romans is basically a dissertation on Christianity; and his other letters include what he claimed the scriptures provided: *"teaching, rebuking, correcting, and training in righteousness, so that every man of God may be thoroughly equipped for every good work."* Indeed, Paul's letters rebuked groups who were out of line, encouraged and instructed others, and had a common theme of producing wise and effective followers of Christ. Paul is uniquely qualified as a scholar and as one who challenged his own **EBS** [ego, bias and self-centeredness] to recognize who Jesus was and how the ancient Tanakh outlines God's plan for all mankind. Letters from other key figures and John's Book of Revelation are also part of the New Testament. Applying literary integrity is crucial in arriving at an appropriate ideology.

When one encounters a seeming contradiction in Scripture or the New Testament, it is wise to pay closer attention to it and evaluate it as opposed to throwing the idea out. This is because some seeming contradictions hold the key to important discovery once they are carefully investigated. One seeming contradiction was recorded in Matthew's Gospel. Jesus had been beaten and mocked by soldiers. It was reported they had encircled him and cast lots for his clothing. He was ultimately hung on a cross, where his hands and feet were pierced with nails. Matthew records Jesus calling out to God. *"My God, my God, why have you forsaken me?"* (Matt 27.46)

This is a strange thing to write, as it appears to suggest that Jesus is a fraud. However, it leads those interested to Psalm 22 written by David 1000 years earlier and recorded in the Tanakh. The Psalm begins, *"My God, my God, why have you forsaken me?"* Psalm 22 goes on to describe events in the final moments of the life of the Messiah 1000 years later – not in the life of David, himself, as none of these things happened to David. It mentions that *"a band of evil men has encircled me, they have pierced my hands and feet. I can count all my bones; people stare and gloat over me. They divide my garments among them and cast lots for my clothing."* (Psalm 22.16 – 18)

Paul, in his writings to the Galatians, then brings us full circle through prophecies and declarations in the Law to explain how God was able to curse Jesus with the wrongdoings of all mankind. This will be discussed in detail later.

There is no good reason why a spiritual ideology requires one to dummy down. On the contrary, when science, history, logic, and a holy book – influenced by a source beyond our space-time dimension – are in harmony, there is a peace that passes all understanding.

Though many of us know people who are fantastic ambassadors for Christ, many inside and outside religious circles are aware of some disturbing ideas promoted by some churches and some professing Christians. Therefore, with accountability to the Tanakh, the Apostle Paul and history and science, the following is a list of six common errant beliefs that need to be addressed. It is no exaggeration to say that millions of lives depend on correcting these reality deviations.

Six Errant Beliefs That Need to Be Addressed:

1 The belief the Earth is less than 10,000 years old

2 The belief there was a global flood – not a local one – and that God instructed Noah to take two of every animal aboard an ark

3 The belief that God <u>reacted</u> to Adam's fall and only then devised a plan to come to Earth as a man and die for our sins

4 The belief those outside of Christianity cannot enter heaven

5 The belief that Christians should aggressively confront and shun the LGBTQ community

6 The belief that most American Muslims are in favor

of radical terrorism to bring down the West

1 The belief Earth is less than 10,000 years old
  Multiple scientific factors point to a universe that is more than 13.7 billion years old. Scripture [the Tanakh] does not provide a complete account of when things occurred or exactly how old the Earth and universe are. However, the Tanakh provides some key indicators suggesting the Earth is much older than 10,000 years.

  Peleg is a descendent mentioned in Genesis 10:25 and that *"in his time the earth was divided."* A little more than 10,000 years ago, rising ocean levels – from the warming of the planet after the latest ice age – covered the land bridges connecting continents. Thus, some of our continents were again isolated. The Earth was again divided. This calibrates a moment in time in the genealogy of descendants of Noah. Peleg lived a little more than 10,000 years ago. Hence, the Bible does not claim the Earth to be 6000 or even 10,000 years old. More on this and the emergence of homo sapiens in Chapter 9.

  Genesis begins with a beautifully poetic description of the six days of creation. However, the Hebrew word translated day is the word *yowm* [or *yom*]. It can also mean a long period of time like an age of time.[1] There is an obvious contextual clue in scripture that *yom* is a long period of time and not 24-hour days. The Jewish day begins at sundown. In Genesis chapter one *"there was evening and there was morning"* is used when describing each of the first six *yom* [days] of creation. But not so for day 7 at the beginning of Chapter 2, where we are told that God rested from his creation [not rested for good]. There is no mention of evening and morning for day 7. This would imply that we are still in the seventh *yom* – period of time.

  A very cool thing happens if we read Chapter 1 of Genesis with the possible interpretation that each day is a long period of time. The passage in Genesis could then be describing the big bang and development of our universe that has occurred over 13.7 billion years. [See Chapter 7 on Time and a Holy Book for a thorough discussion of Genesis describing the big bang and the correct progression of cloud cover breaking up to eventually reveal the sun and stars.] Scripture and science agree here. Teaching a young Earth – as many Christians do - is extremely detrimental. This implies to these conservative Christians that scientific discoveries - different from scientific theories - cannot be trusted.

It is detrimental and potentially dangerous when a religious organization teaches that we should ignore or disrespect scientific discovery and then teaches doctrine that contradicts scripture and history. When people feel justified mistrusting science, they leave themselves open to influence from misinformation. Assumptions made during the COVID-19 pandemic and with regard to critical issues like climate change underscore the consequences of such mindsets.

2 The belief there was a global flood – not a local one –
and that God instructed Noah to take two of every animal aboard an ark

The Tanakh [Bible] does not describe a global flood. It describes a localized flood – likely, the entire Mesopotamian Plain. And it is not true that God instructed Noah to take two of every animal. Chapter 7 of Genesis is where the account of the flood begins. In verses 2 & 3 God told Noah to take 7 pairs (a male and female) of certain animals, 2 pairs of others, 7 pairs of birds, and by implication none of other animals. A look at the meanings of the Hebrew words used helps determine that only certain types of animals were needed aboard the ark. Livestock for when Noah's family started over, animals for sacrifice, and birds were among the necessary animals on board. The majority of animals around the globe were likely minimally affected or unaffected by this flood. This is because the flood was not global.

In Psalm 104, there is a poetic description of creation and the progression of Earth. It declares that, after the land masses came up out of the water initially on Earth, God would never again allow the tops of the mountains to be covered with water. Thus, a global flood in Noah's day would contradict this scripture.

As mentioned, the Genesis account of the flood yields context that the flood was not global. Genesis 8: 5 – 6 describes how, after the mountain tops were already visible, the dove could find no place to set its feet, *"because there was water over all the surface of the earth."* The Hebrew word for earth can have various meanings. The context in this case implies a region or area of habitability due to knowing that mountain tops were no longer covered with water. Combining what is and is not possible according to scripture with what archaeology and other branches of science reveal produces a few scenarios of what may have transpired with humans over the past 200,000 years. These are discussed later in this book.

The acceptance of a *global flood* among Christians has led to the acceptance of more misinformation. Non-scientific and non-scriptural ideas exist in some Christian circles of how a massive flood a few thousand years ago makes a young Earth appear old.

3 The belief that God <u>reacted</u> to Adam's fall and only then devised a plan to come to Earth as a man and die for our sins

As one grows in the ability and commitment to interpret scripture with accountability to science and history, one learns to live with possible scenarios while gleaning the important points.

Regarding the Tanakh and New Testament, there is a huge problem with many Christians who teach that this plan was only devised after

Adam sinned – after humans displayed the propensity to mess up. The biggest problem with a Creator existing beyond our space-time dimension and <u>reacting</u> to the propensity for humans to sin is the suggestion that God can be fooled and is not in control.

Genesis 1:1 references *In the beginning*. This implies that our space-time dimension (our universe) had a beginning [out of nothing that can be seen]. This is scientifically accurate. Thus, God exists outside of our space-time dimension. Genesis mentions that God declared *"Let us make man in our image"* (Gen. 1.26), establishing multiple persons of the godhead - namely the Father, the Son and the Holy Spirit. The Apostle Paul tells us in 2Timothy 1:9 that the grace given us in Christ Jesus was given to us *before time began*. Thus, Paul implies that the plan for the Son to redeem us through his sacrifice was established before our universe began 13.7 billion years ago.

Because God exists in such a realm outside our space-time dimension, He exists in our past, present and future simultaneously – *the same yesterday, today and forever.* This is how God is in control while giving us a free will. And thus, God is never fooled. He sees and is *in the future.* This notion that he initially constructed this universe to be eternal but then Adam sinned and messed up everything is not what Paul teaches.

Accepting that God *reacted* to man's fall and did not anticipate it leads to many contradictions to reality, scripture, God, and the plan. Additionally, the laws of Physics suggest this universe is finite. Left to itself, it will eventually cease to exist. This universe was not intended to be eternal. Again, scripture and science agree. God works on an eternal habitat for us in another space-time dimension.

One of the most important recognitions of God existing outside of our space-time dimension and knowing what will happen when He creates is that He created beings He knew would rebel and become *"the spiritual forces of evil in the heavenly realms"* threatening our lives. (Ephesians 6: 12) If God is in control as He declares, then He knew the evils He was creating. They must have purpose in this life as it relates to the life to come. Such context helps to better understand the impact of pain and suffering in this beautiful but often brutal first life. This will be discussed in Chapter 8. If on the other hand God is not responsible for the evil in this life and was fooled by the fall of man and only then devised a plan to redeem us, then God would not be in control. In such a scenario there would be little reason to trust His promises.

We have become a society of people that seem to be okay with fanciful spiritual perspectives that can go in any direction without a requirement of accountability. This leads to critical contradictions with no clear, supported ideology. Today, many assume they are correct with little recognition of their lack of accountability. People despise others for their beliefs, argue over beliefs, refrain from discussing beliefs and mock spirituality. This is all very needless, illogical and detrimental; but it is often how we humans behave.

4 The belief those outside of Christianity cannot enter heaven

We will confirm the validity of the covenants established through Abraham and Jesus. The signs and plan are quite clear. But Jesus said something that appears to be exclusionary; and it is where many get the notion that no one else will be allowed into heaven other than those who have made a commitment to Jesus in this lifetime. Jesus said, *"I am the way, the truth and the life. No one comes to the father except through me."* (John. 14.6) This seems straightforward and exclusionary, but it is not.

Consider a sincere question posed by many conscientious people: "What happens to people in remote parts of the world who never hear about the Messiah? Can they go to the afterlife - heaven?" Thankfully, Paul gave us the answer in Romans 2: 12 – 16. Again, it is challenging to understand all major concepts surrounding the Law. It is why an entire chapter is devoted to it later. The contrast of the letter of the law versus the spirit of the law very much applies to what Paul teaches.

"All who sin apart from the law will also perish apart from the law, and all who sin under the law, will be judged by the law. For it is not those who hear the law who are righteous in God's sight, but it is those who obey the law who will be declared righteous. (Indeed, when Gentiles, who do not have the law, do by nature things required by the law, they are a law for themselves, even though they do not have the law, since they show that the requirements of the law are written on their hearts, their consciences also bearing witness, and their thoughts now accusing, now even defending them.)

And just to be clear that Paul is indeed speaking of the judgment and the afterlife, the passage finishes with this:

This will take place on the day when God will judge men's secrets through Jesus Christ, as my gospel declares." (Rom. 2.16)

Paul is saying that those who do not hear the gospel will be judged by how they follow the leading of their heart during their lifetime. We will see through the Law and from before time began that the beating and death of God himself – the Son – is the gateway to the afterlife. Just as priests were taught to sprinkle the blood of an atonement sacrifice on the mercy seat in the tabernacle on Earth, the blood of God the Son would be sprinkled on the mercy seat in heaven. It is only through this act of grace that anyone is allowed into the afterlife prepared by God - including the ones who never hear the gospel or the Law.

Paul suggests they will be accepted due to faithfully following their heart (and not knowing to put faith in Jesus specifically). In other words, many Buddhists in Southeast Asia, Hindus in India, and Muslims in the Middle East will be welcomed into the afterlife because of how they followed the leading of their heart. Thankfully, God will do this in many instances. And the payment allowing this will be the blood shed by the co-Creator of this universe according to the plan devised before time began and spelled out in the Tanakh.

Will people of many faiths and ideologies enter into the afterlife? Yes. How will it be possible? By the crucifixion of God's Messiah, prophesied centuries before the arrival of Jesus. So whether one is a Christian or one belongs to another faith and lacked sufficient opportunity to learn of the covenant with Abraham and its completion through Christ but followed the positive leading of their heart, the way to the Father is only through Jesus. Hence, the accuracy and inclusion that Jesus is *"the Way, the Truth and the Life."* (John. 14.6) No one will come to the Father except through him.

Some Christians argue vehemently against this notion that non-Christians will be allowed into heaven – citing Romans 10: 9, 10, where Paul states that we are saved by confessing with our mouth that *"Jesus is Lord."* The verse does not logically imply that confessing Jesus is the only way. After all, most of us would agree that Moses, David, Isaiah and Daniel will be in heaven. None of them confessed Jesus as Lord. They lived centuries before the crucifixion. They did not know him by name. If confessing Jesus is Lord is the only way to heaven, then Moses, David, Isaiah, and Daniel will not join us in the afterlife. Because we are assured Moses and David and the others are in heaven in another space-time dimension, confessing Jesus is not the only way to the afterlife.

An important distinction should be considered. At this time, most of the world's population knows of Jesus and has an opportunity to investigate claims surrounding him. For instance, no one reading this book can argue they never had a chance to consider Jesus as the Messiah. Arguing admittance to the afterlife by following their heart would not apply. Thankfully, God is perfectly fair. It is not our job to judge. It is our job to share the plan and offer evidence to support it.

5 The belief that Christians should aggressively confront and shun the LGBTQ community

Most people in the LGBTQ community, like most of the rest of society, are very well adjusted. Many could care less how others treat them and perceive them. Over years of teaching, I got to know many LGBTQ students with beautiful hearts. But I learned how poorly some were treated – especially by Christians. One of my students, who also attended a Bible and Science group I led at the home of a family I knew through school and church, shared her experience. She informed her parents she was gay two years earlier. They were Christian. Immediately her cousins, with whom she had a life-long loving relationship, were instructed never to speak to her again. Her parents looked down on her and explained how she must change. She was hospitalized twice in the next six months for depression and suicide.

Many of my Christian acquaintances speak of gays with comments like, "God loves the sinner but hates the sin." But it appears to come from hearts that believe gays are far greater sinners than themselves – so bad, in fact, that we should force them into sex therapy. We would not dream of doing this to the significant percentage of American Christians that are obese. But somehow it is okay to look down our noses at gays and insert our will into their lives.

One of the simplest definitions of love is to believe the best in people. While some claim there is proof that no one is born gay, this would be nearly impossible to conclude via observational study. If Christians need a verse to connect the notion that, over centuries, the evolution and consequences of humans could lead to people being born gay or lesbian, Romans 8: 19 – 23 may be appropriate. Notice in the passage that God himself subjected creation to frustration.

"The creation waits in eager expectation for the sons of God to be revealed. For the creation was subjected to frustration, not by its own choice, but by the will of the one who subjected it, in hope that the

creation itself would be liberated from its bondage to decay and brought into the glorious freedom of the children of God.

We know that the whole creation has been groaning as in the pains of childbirth right up to the present time. Not only so, but we ourselves, who have the firstfruits of the Spirit, groan inwardly as we wait eagerly for our adoptions as sons, the redemption of our bodies."

Here is a powerful question: if God showed up and told us some persons are born gay, how would it affect the way we treat people in the LGBTQ community? Why would we assume being born gay is <u>not</u> a possibility? Some religious persons are afraid to change their mindset and behavior to a more loving and faithful perspective. Generations of teaching and pressure may leave them feeling they would be turning their backs on God to believe persons can be born gay.

When white Christian children are raised by spoken and unspoken actions portraying certain people groups as beneath them, it explains much of today's behavior. Some of my former black and Latino students shared that they transferred colleges because many white conservative Christians at a particular university treated the minorities as inferior or as charity cases through verbal and non-verbal signals.

Chapter one of Paul's letter to the Romans is often quoted by those who want to outwardly condemn all homosexuals. However, such an interpretation misses context. And the greater message continued into Chapter 2 is that none of us should judge harshly anyway. To ignore our own humanity and to miss the point being made by the Apostle Paul is what he warns of in the second Chapter of Romans where he states that *"God's name is blasphemed among the Gentiles because of you."* (Rom. 2.24)

With regard to sexual sin, most of us can recognize perversion when we see it - the (child) pornography industry, human trafficking, incest. Some

31

assume the moral of the story regarding Sodom and Gomorrah is that homosexuality is what we all must defend against. But from a logic and math standpoint, this is not a valid conclusion. The explanatory variable [root cause] was runaway lust. The response variable [the result or action from runaway lust] was outward displays of immorality of both a heterosexual and homosexual nature. And most of the laws given to Moses involving lust and immorality pertain to heterosexuality – not homosexuality. Homosexuals did not corner the market on sexual sin. Yet I have listened to Christians I respect say that homosexuality is the single greatest threat to America.

The primary text in the Christian New Testament involving homosexuality is found in Chapter 1 of Romans. In Romans 1: 18 - 32, Paul is addressing *"men who suppress the truth by their wickedness, since what may be known about God is plain to them."* He is not addressing gays per se. Continuing his thought, Paul begins verse 21 with *"For although **they** knew God, **they** neither glorified him as God nor gave thanks to him, but their thinking became futile and their foolish hearts were darkened."* 1 Samuel 16: 7 reminds us that *"People look at the outward appearance, but the Lord looks at the heart."* God knows **they** are men and women with *darkened hearts*. These are people who knew God exists and strayed far and with horrible mental processing and depravity in their hearts.

As Paul continues, he is obviously speaking of the progression of sin in humans over time as a result of the suppression of truth by wickedness. Verse 23 mentions how men made idols in the forms of birds and animals and reptiles. In verse 24 he mentions sexual impurity [in general]. Verse 26 and 27 mention that God gave them [men who suppress the truth by their wickedness] over to *"shameful lusts. Even their women exchanged natural relations for unnatural ones. In the same way the men also abandoned natural relations with women and were inflamed with lust for one another. Men committed indecent acts with other men and received in themselves the due penalty for their perversion."*

Today, there are heterosexual men who may or may not express faith in Christ. They may dabble in pornography or have many partners. As these men proceed further into sexual activity, it may lead to *"indecent acts with other men [plural]."* Some may contract STDs. This type behavior matches the tone of this passage. It does not appear to describe heterosexuals and homosexuals who are monogamous and generally under control. But for those who want to condemn all gays, it is important to continue following Paul's discussion of the progression of sin, and all the abhorrent human acts he mentions. Here is the passage immediately following verse 27:

*"Furthermore, since **they** did not think it worthwhile to retain the knowledge of God, he gave them over to a depraved mind, to do what ought not to be done. **They** have become filled with every kind of wickedness, evil, greed, and depravity. **They** are full of envy, murder, strife, deceit and malice. **They** are gossips, slanderers, God-haters, insolent, arrogant and boastful; **they** invent ways of doing evil; **they** disobey their parents; **they** are senseless, faithless, heartless, ruthless. Although **they** know God's righteous decree that those who do such things deserve death, **they** not only continue to do these very things but also approve of those who practice them."*

Of the many sins Paul described in Chapter 1, millions around the world have focused on homosexuality as the unpardonable sin. Why not gossips or slanderers? Thankfully, it does not matter much; because Paul continued his teaching,

"You, therefore, have no excuse, you who pass judgment on someone else, for at whatever point you judge the other, you are condemning yourself, because you who pass judgment do the same things. Now we know that God's judgment against those who do such things is based on truth. So when you, a mere man, pass judgment on them and yet do the same things, do you think you will escape God's judgment? Or do you

show contempt for the riches of his kindness, tolerance and patience, not realizing that God's kindness leads you toward repentance?" (Rom. 2. 1 – 4)

A necessary word of instruction: Paul is not saying that each of us has committed each of these sins or to the greatest extent. Paul is simply being consistent with his teaching of the purpose of the Law regarding the Christian faith. Each of us is to recognize that we fall short of perfection. Self-righteousness has no place.

This is huge. Paul has embarked on decoding the judgment and forgiveness God gives. In other words, Paul is beginning his dissertation on the Christian faith. But too many people do not understand this. This lengthy letter to the Romans – in addition to his other letters in the New Testament - thoroughly explains the connection of the Law to God's plan and to our sin, to God's wrath and judgment, to God's grace and kindness, and to provisions for those who never hear the message. All these nuggets of wisdom, written by perhaps the most qualified scholar to share them, are contained in the 16 chapters of Romans. But tens of millions of today's professing Christians have lost the message early in Chapter 2, because they don't choose literary integrity over peer pressure. They get lost in self-righteousness.

There are many reasons why it is important to address this issue. Many people of integrity will not even consider Christianity due to the treatment of gays. Prominent, respected evangelists have insisted that Christians confront the LGBTQ community aggressively. Just like Paul rebuked Peter for straying from accurate doctrine, Paul would rebuke many of these prominent men and women of God today for their campaign against people within the LGBTQ community.

6 The belief that most American Muslims are in favor of radical terrorism to bring down the West

Abraham was Ishmael's father. This is basically the origin of Arab people groups, of which many today are Muslim. God responded to Abraham's request for Ishmael to live under God's blessing:

"And as for Ishmael, I have heard you: I will surely bless him; I will make him fruitful and will greatly increase his numbers. He will be the father of twelve rulers, and I will make him into a great nation." [Genesis 17: 20]

The recent disrespect and mistrust of most American Muslims by millions of conservative Christians is disturbing. To assume that any religious group is defined by its extremists is illogical, and many people tend to over-simplify what it means to be Muslim. Those of us with Muslim friends and acquaintances find it disheartening that so many accept the stereotyping of Muslims. Muslims, Jews, Christians along with many other groups agree of the need to battle against Islamic radicals and Christian radicals – especially anyone willing to take another life in an act of terror. We can agree that those groups must be stopped. They murder innocent men, women and children. They do not honor the God of Abraham, and their vision of God's will on Earth is disgraceful to all the Abrahamic faiths and to humanity.

In my years of teaching, some of the most conscientious students I taught were Muslim. One attended the Islamic Center in our area and shared powerful teachings of love and service to others from the Imam there. She wanted to choose a major in college that would provide the opportunity to care for those less fortunate. Her joy and kindness to others of all faiths and cultures naturally endeared her to many. Another Muslim student attended the classes I led on Science and Bible.

Those who condemn all Muslims as extremists provide some of the fuel for the radicalization of individuals around the globe. Some of my friends and relatives who forward emails and information condemning Muslims seem to lack the accountability of their role in promoting misinformation. Ironically, one of the greatest examples of a culture respecting other beliefs and inviting persons from various cultures to collaborate to gain knowledge and understanding was the House of Wisdom in the Golden Age of Islam.

Today some compassionate scientists make loving attempts to help some Christians and followers of Islam whittle away at bigotry and racism and develop respect for differing views. One scientist I respect, describes people having *"different realities."* I wonder if it may be wiser to call them different *paradigms.* It seems to me there is great danger when an ideology is **assumed to be reality**. When this occurs, the allegiance is to the assumed reality. Such allegiance may be elevated above objective evidence to the contrary and can produce an over-confidence or certainty that can lead to dogmatic, negative, and dangerous actions. Respecting and considering infers differing interpretations of life. None of us have a perfect grasp of *reality* – the exact ideology in line with science, history, and human behavior. Using the term *paradigm* does not give the right of anyone to absolutely define reality. Instead, it speaks to a necessary degree of humility as we journey toward reality.

Discussing differing views civilly is not easy but is very necessary. Most of us lack the grace and understanding of how to interact with others of differing views. Many times I have missed my own underlying tones and imperfections in communication. But one can learn much from people who hold different beliefs. We can disagree and feel passion about the afterlife and how it is achieved. And ultimately it is each of our responsibilities to seek the right path and hold ourselves accountable.

One final thought – particularly to those who have been negatively affected by Christianity. Paul and Jesus turn religiosity and self-righteousness on their heads. The compassion and love Jesus displayed in his time on Earth reminds us of this love God has for us. Jesus, when accurately defined, is very much who most hope God to be. People who promote an inaccurate definition of Jesus and an inaccurate Christian message can inflict tremendous pain amplified by guilt and shame. This is not a minor issue. On a micro level, it can lead to depression and suicide. On a macro level, it can lead to social injustice and even to genocide. Yet a strange recognition is that a God truly in control is allowing such inaccuracy and such trauma to occur.

Chapter 4

Science: The Known and The Unknown

I asked my AP Statistics students to estimate the percentage of astrophysics knowledge known and understood by Stephen Hawking. I then waited for them to share estimates aloud.

"30 percent. 40 percent. 25 percent."

"Before we give a better estimate, look at this tennis ball I have in my hand," I said. "The volume of this tennis ball represents the amount of all knowledge and information known in the world 36 years ago when I graduated from high school. Medicine, economics, behavioral and physical science, ... you get the idea. So what volume represents the total amount of knowledge and information today?"

"The answer: covering the entire United States with tennis balls to a height of two feet. Ten years from now, it should be approximately the volume of the entire planet Earth. A tennis ball 36 years ago. The size of Earth ten years from now. If you could read thousands of books per second, you still could not keep up. Give me a better estimate of the percentage of astrophysics knowledge known and understood by Stephen Hawking."

After a few moments of class discussion, we recognized it would be far less than one percent!

Sixty years ago, most scientists thought there must be billions of planets with advanced animal life in the universe. They thought a primordial soup spawned life on Earth and that self-replicating DNA emerged easily and quickly. 36 years ago some scientists were questioning the notion of a primordial soup and how self-replicating material came to exist. Remember the illustration of the tennis ball. Today, we have a trillion times the amount of information and knowledge compared to 36 years ago. Researchers must specialize in narrowly defined fields of science, because there is so much to analyze and discover. The assumptions made about the existence of millions or billions of planets with advanced animal life should no longer be assumed due to the wealth of discoveries made in the past few decades.

With these things in mind, here is a statement that is contrary to what most learned in school or witness in some documentaries. It may seem like it shouldn't matter, when it is quite critical. The statement is this: based upon the sum of current research from investigating astrophysicists and chemists and biologists, we would have to conclude that animal life is not at all abundant in our universe. While there could be a fair number of planets capable of *supporting* life if animal life were transported there with the proper additional resources, there is likely not even one other planet in the known universe that would have the hundreds of necessary characteristics similar to Earth's providing it the protection and resources to accommodate four billion years of preparation and evolution to produce advanced animal life such as we have here unless something from outside of our space-time dimension intervened.

In the realm of research and statistics and the scientific method, a hypothesis must be supported with probability theory to be taken seriously. Some people dismiss probabilities out of sheer ignorance. Others know just enough about probability to be dangerous and continue to promote ideas with great confidence, when in fact a more accurate mathematical and scientific approach would help them to recognize the frailty of their arguments. This happens more frequently than one might

39

expect among people possessing advanced degrees. In particular, it is *conditional probability* that tends to hold the key in supporting a hypothesis or ideology.

Some of the more intelligent misunderstandings of probability include the following. "We are here, so there must be an explanation for why we are here." And then the individual *assumes* that the explanation includes only sources from *within our space-time dimension* with no influence from a being beyond our space-time dimension. Such is the foundation of Naturalism, a philosophy that is not pure science but is what we have taught as science in American public schools for decades.

Another intelligent misunderstanding of probability proceeds something like this: "We know our very existence on Earth defies the odds, but with millions or billions of other planets like Earth in the universe, it is logical to understand that at least one of those chances would produce a planet like Earth with its incredible life forms." With this second erroneous conclusion, many highly intelligent and mathematically inclined individuals excuse probabilities and resort back to the first intelligent misunderstanding – convincing themselves that there is no God, and that it is very logical to conclude the presence and progression of life on Earth occurred via random chance and without the need of an overseer.

Pondering what has been learned about our planet can help produce a realistic perspective from which to analyze. Here is a summary of what has taken place on Earth in the past 4.5 billion years:
Complex life on our planet required self-replicating material like DNA that could produce life and then allow the replication of life. While many assumed this self-replicating material could emerge four billion years ago from substances contained on Earth, the evidence points to a different conclusion. Protocells have been touted by some as a potential intermediary step between prebiotics and origin of life; but self-

replication is necessary to produce protocells.[1] So while protocells may be an intermediate step, they do not explain the origin of self-replicating material.

Some biologists who do not grasp this point. They respond with statements like, "It is easy to copy and reproduce DNA." But they may not take the time to realize it is easy to do such a thing *today*, because *today* we have DNA. With it already in existence, we can easily copy it. We are speaking of the conundrum regarding how Earth first produced DNA or any other self-replicating material four billion years ago. The one who can answer that question and support it may well receive a Nobel Prize. This is because we do not yet have an answer.

Many researchers are hard at work trying to determine the origin of nucleotide bases and of self-replicating material. In 2016 NASA scientists reported that simulated conditions of ice samples on meteors travelling in space at -440 degrees Fahrenheit can, when subjected to UV rays, break down and form into uracil, thymine and cytosine.[2] DNA strands are formed with pairs of bases. Adenine pairs with thymine, while cytosine pairs with guanine. In RNA, uracil replaces thymine. The problem is that uracil, thymine, and cytosine form under these extremely cold conditions. Their corresponding base pairs, guanine and adenine, do not.[3] Additionally, Pearce and Pudritz [2015] found that cytosine fails to survive within meteorite parent bodies as a result of aqueous deamination; and thymine does not survive when H_2O_2 is present because of an oxidation reaction between the two.[4]

One of the reasons thermal vents in the ocean were once thought to create the extreme hot and cold necessary to form the base pairs, is because of the presence of cold and heat. However, the cold of an ocean is far from -440 degrees Fahrenheit. As of this book's publication, there is no substantiated answer for how nucleotides emerged and paired to make RNA and DNA possible.

There is a plethora of elements on our planet, but initially a vast amount of these were water-soluble. This means that water on Earth nearly four billion years ago was poisoned with high levels of lead, zinc, copper and other elements. Prokaryotic bacteria needed more than a billion years to convert these water-soluble elements to non-water-soluble and to generate an ever-increasing volume of atmospheric oxygen.[5] A huge and beneficial consequence of the bacteria accomplishing this is that vast deposits of iron ore and copper would be available billions of years later when homo sapiens arrived, so that we could eventually create the technology we enjoy today.

A billion years ago more complex life forms emerged in addition to bacteria, including sponges and worms and other larger species wiped out by a mass extinction prior to the Cambrian explosion. Four more nicely-timed and beneficial mass extinctions, each followed by swift repopulations and ecosystems, have occurred since the Cambrian.[6] Those events helped provide a large supply of fossil fuels waiting for humans at just the right time.

Additionally, given the sun's changing luminosity, had it not been for these mass extinctions and repopulations at ideal times, there existed the possibility that Earth might otherwise have been permanently sterilized.[7] Perhaps the five major extinction events are rarely discussed in schools due to the problems they present in attempting to promote a smooth evolutionary model embraced for decades now.

The emergence of the new ecosystems – these explosions of life – behave as if *anticipating* the soon-to-occur atmospheric and solar changes – something that makes more sense if a God orchestrated the change. Natural change tends to be *reactionary* rather than anticipatory, which is why these more *anticipatory* extinctions and repopulations lend credibility to influence from a source beyond our space-time dimension as opposed to strictly terrific luck and timing if only nature is responsible.

For a planet like Earth to have the billions of years to spawn, develop, and nurture life, it needed protection from harmful rays that would otherwise sterilize the planet and cause it to become barren. This requires a sustained electromagnetic field. In order to accomplish this, radioactive elements are necessary. The decay of radioisotopes within our planet maintains a sufficient level of heat in our core to keep iron and nickel molten and spinning in order to maintain the electromagnetic field. Radioactive isotopes also generate tectonic activity required to sustain life.

Water is not rare in the universe. But Earth's proportional composition of elements, minerals and water are quite rare and quite necessary for our planet to support and promote life.[8] It is not until the nine billion-year mark of the universe that the abundance of radioactive isotopes and other heavy elements are sufficiently present in the universe along with key elements like flourine.[9] Thus, Earth formed at a nearly ideal time in the universe for acquiring the heavier elements it possesses. It is not that Earth-like planets were impossible to form before the nine billion-year mark. It is that it was far less likely. And arguably, without all these elements finally present at the nine billion-year mark, animal life might be impossible or nearly impossible.[10]

For our sun and its forming planets to grab these elements, our solar system must have formed in a more dense location of the galaxy, where enough explosions of various stars provided elements like fluorine in addition to radioisotopes and the other heavy elements.[11] However, we could not stay in that same location with so many forces at work. We needed to end up between the arms of the galaxy – in our galaxy's habitable zone.[12] Hence, it is not just that the plethora of heavy elements were finally available in the universe. It is that a planet and its system formed in a location where it could grab all these elements and then moved to an area without so many rays and forces at work that would otherwise prevent animal life from developing.

Earth again defies the odds with a high mass density and composition of heavy metals. Typically, this occurs in planets closer to the sun but not in the habitable zone of the solar system. Despite Mercury being the planet much closer to the sun, it is Earth possessing the greater mass density.[13] In addition, Earth is far enough from the sun to be in its habitable zone. We seem to catch every lucky break.

Earth needed correct timings of increases and decreases in certain minerals and gases. The Great Oxygenation Event 2.4 billion years ago was a result of a nickel famine that led to a significant decrease in methane, that in turn allowed for an increase in atmospheric oxygen.[14] Yet, if this chronology of events had occurred earlier, when the sun's luminosity was lower, Earth could have transitioned into a long-term ice ball.[15]

Mathematically, the timings of so many key interactions over four billion years continues to signal the rarity of a planet able to coordinate an array of factors in order to produce the necessary progression toward more sophisticated life. It is as if a God knew what conditions – like the sun's luminosity – were right around the corner and intervened with mineral and gas interactions, mass extinction events followed quickly by new ecosystems, and the ideal time necessary to have optimal amounts of natural resources for an advanced technological civilization 4.5 billion years after initially forming.

Animal life is not at all common in this universe. Too many things need to occur and interact in a just-right fashion.
In 1961, Dr. Frank Drake introduced his Drake equations, with which he attempted to estimate the number of advanced civilizations in our galaxy.[16] Based upon the scientific evidence of his day, there was no

reason to doubt his conclusions that billions of civilizations should exist in our galaxy alone. In the 1970's and 1980's, Carl Sagan also proposed that there must be many civilizations within the universe with "billions and billions" of stars out there. Sagan was more than an excellent scientist. He was an outstanding human being. It is tempting for us to hold to his beliefs because of his appeal. However, subsequent scientific discovery no longer supports his contention.

In 2016, NASA reported that it found an additional 1284 exoplanets. But only nine were adequately rocky and the right size and in the habitable zone.[17] That is less than 1%. And there is a difference between a planet that is capable of "supporting life," if existing life was transported there, and a planet capable of spawning, nurturing and producing advanced, technological life such as what Earth accomplished.

Mike Wallace, Space.com senior writer, quoted NASA scientists as saying the following, "About 550 of the newly validated 1284 [exoplanets] are small enough that they could be rocky, scientists said. And nine of those might be able to support life."[18] Again, the ability of a planet to *support* life is much different from a planet that can spawn and nurture life. Balanced information like this should be shared in colleges and high schools. But such information rarely finds its way into the classroom. I will never forget the advanced student who shared her assumption that, because there are habitable zones, there must be much animal life in the universe. True scientific pursuit helps us recognize the need for much more than merely a small rocky planet, with some surface water, abiding within a habitable zone of a solar system and galaxy.

Consider Earth's need for an exceptionally large moon. Our moon is literally one-in-a-trillion. There is still some uncertainty as to how Earth captured such a large, ideal moon with its gravitational force. The most likely natural explanation would be a collision with our moon as both were forming.[19] We have an ideal 23.5-degree tilt-of-axis to our sun and an ideal spin rate today as a result. This tilt gives us a much milder climate in a far greater swath of our planet. The fact that our moon is so massive relative to Earth means that it is able to create sufficient tidal activity to cleanse ocean waters from toxins and replenish them with nutrients; and it has the gravitational force to hold us at our ideal tilt and not allow the sun's gravity to yank us more upright and dramatically slow our spin rate – which would lead to harmful climate changes.[20]

Our spin rate is very favorable to life. Spin too slow and experience a daily temperature variance that is too great – leading to a destructive impact on plant and thus animal life.[21] At one point in the distant past when Earth spun too quickly, winds at times likely approached 400 miles per hour. Humans arrived at the time of ideal spin rate of Earth. Thankfully and coincidentally, humans arrived at the ideal time regarding the steady burning state of the sun and the greatest amount of useful fossil fuels. In the future, natural gas would convert to less beneficial methane gas.

We would not have expected to capture such an ideally large moon in our gravitational pull had it not been for a precise collision.[22] Finding other planets with sufficiently large moons relative to the size of the planet has an extremely low probability. A planet needs a large moon like ours to be a candidate for lower life forms evolving to higher life forms.

As our age of discovery emerged, we recognized another coincidental benefit from this moon. It just happens that our moon is 1/400th the diameter of the sun and 1/400th the distance from Earth to the sun. This allows for occasional total eclipses – the moon nearly perfectly blocks the sun and allows scientists to analyze. The observations made by scientists during total eclipses of the past 200 years have revealed the chemical make-up of bodies in space, confirmed Einstein's theory of general relativity, and allowed us to learn much regarding the universe while standing firmly on our planet. The incredibly coincidental ratio of the diameter of the moon to the diameter of the sun and the ratio of the distance of the moon and sun from Earth has allowed this to happen. It must be extremely rare to have these identical ratios of distance and diameter involving a planet, its moon, and its sun.

One of the easiest requirements to comprehend regarding protection of a developing planet for billions of years is the need to have four giant planets in the solar system whose gravitational pull can divert asteroids and comets away from colliding with Earth. No matter where we are in our orbit of the sun, Jupiter, Saturn, Uranus and/or Neptune are protecting us by diverting huge space debris. Without at least four giant planets in the same solar system, a planet could not accomplish what Earth has accomplished. The asteroid strikes or collisions with comets would be far too frequent and too severe to allow the billions of years necessary for life to progress. In 1994 fragments of a comet were pulled into Jupiter with significant force.[23] The resulting explosion rose 10,000 miles above Jupiter's surface. There is always at least one of these giant planets that has our back. Also, our tiny blue planet is fortunate to have the asteroid belt contained where it is.

In addition to requirements most of us would understand, there are many more requirements being discovered by astrophysicists and chemists in specialized areas. These include things like "the infall velocity of matter into the dark matter halo of the potential life support galaxy," and "the average rate of migration of aqueous fluids through the planet's upper crust." These and more than 900 other documented requirements can be found in the appendix of the book, More Than a Theory by Hugh Ross [2009]. Ross also estimates and posts probabilities for each characteristic listed.

As persons ponder whether animal life on Earth is a result of purely natural factors or if there has been influence from a source beyond our space-time dimension, it is important not to reach unsupported conclusions. Just because science does not currently have explanations for things like the origin of self-replicating material does not imply that it must be a God who designed it. Similarly, it doesn't *imply* a natural explanation exists. While it is tempting to reach unsupported conclusions, one must accept uncertainty and the role of faith in both science and religion.

Wide ranges of probabilities are given for the likelihood of the known universe producing a planet like Earth possessing the characteristics required for advanced animal life. What follows is a conservative estimate. It is not appropriate to multiply all probabilities of the more than 900 characteristics. This is because many probabilities are dependent with one another. Some events may cause two or three characteristics to occur simultaneously.

For example, suppose characteristic A has a .001 chance of happening. Event B has a .1 chance of happening. And characteristic C has a .8 chance of happening. Suppose also that each characteristic depends on the same event occurring in space. Multiplying the three probabilities would suggest a .00008 chance of these three characteristics occurring near an Earth-like planet. It implies happening only 8 times per 100,000 planets. But due to dependency, the accurate probability is some value between the least likely individual characteristic of .001 and the product of the three probabilities of .00008.

There are approximately 10^{22} stars in our known universe. Most stars have no planets due to their locations among dense star clusters. Therefore, if the probability of a planet forming naturally with all the necessary life characteristics is .000000000000000000001, we would expect only one such planet in our universe. This is a decimal probability with 21 zeroes followed by a non-zero digit. Taking dependence of characteristics into consideration, the actual probability must be much

more dramatic. Reasonably, it should be at least as extreme as this value:
.000000000000000000000000000000000000000000000001

This is a result of considering the many discoveries made by astrophysicists and chemists who are researching many narrowly defined features of the universe and life. The conservative probability shown above would require a trillion trillion universes before another planet formed with the required characteristics to spawn and nurture life. [A trillion trillion is this: 1,000,000,000,000,000,000,000,000.] The decades-old notion that billions of planets in our known universe have animal life should be updated and held accountable to the vast amounts of information and insight provided by respected scientists and institutions.

Conditional probability helps us to recognize which ideologies and scientific hypotheses are best supported by the evidence of nature. **Given that** there are likely no other planets with the hundreds of necessary requirements for life to begin and be nurtured and then develop into advanced technological civilizations and **given that** there are huge problems and contradictions within purely naturalistic explanations of the progression of life on Earth, and **given that** there is currently no plausible explanation for the chemical production of self-replicating material on Earth prior to the emergence of life, and **given that** there exists a holy book which contains scientific information centuries and millenia before their scientific discovery, and **given that** this holy book contains prophecies and signs that must have come from a being beyond our space-time dimension, the conclusion that a God exists and has had a role in life on planet Earth is far more likely than putting one's faith into strictly random natural occurences getting us to where we are today.

Returning to the statement that "We are here, so there must be an explanation" (but then assuming the explanation must be random chance from within our space-time dimension with no possibility for involvement from a being beyond our space-time dimension) and to the statement, "We know that our very existence on Earth defies the odds, but with millions or billions of other planets like Earth in the universe, it is logical to understand that one of those chances would produce a planet like Earth with its incredible life forms," we now can recognize the errors in both statements.

The second statement is absolutely false. There are certainly not billions of planets in our known universe with all of the fortunate, necessary requirements for advanced life to emerge and be sustained.

The first statement is easily assessed via conditional probability. We are here, and there must be an explanation. We have two primary explanations. One is that random chance explains everything. But this probability is miniscule. The other explanation is influence from a being beyond our space-time dimension. The second explanation is by far the most plausible via occham's razor (the simplest explanation). There exists a holy book accepted still today by over half of the world's population. It was written and compiled between 2500 B.C. and 150 B.C. It describes the big bang beginning of the universe. It mentions God existing outside of our space-time dimension. It describes the continued expansion of the universe. It has clear descriptions of significant historical events hundreds and even thousands of years before they occurred. It has signs and prophecies of a Messiah that have been fulfilled.

Here is a keen tongue-in-cheek statement to consider: **If God intended for us to recognize this reality, wouldn't he have given us reasoning skills and placed our planet at an ideal location in the galaxy to observe the past 13 billion years in space, and given us a moon whose ratio of distance from Earth and distance to the sun are in proportion to the ratio of the diameter of the moon to the diameter of the sun so that we could study and understand so much about the universe while remaining on or around our planet? Wouldn't he have allowed hundreds of scientists to contribute to a list of hundreds of requirements for a planet like Earth to form so that we could recognize that random chance was not a plausible explanation for human life on** our planet? Wouldn't he make it possible for us to observe and grasp the phenominally precise expansion of the universe and the precise balance of forces in the universe? Wouldn't he have allowed us to discover that the building blocks of life did not originate on Earth itself and that we have no known natural source for such prebiotics? Wouldn't he have made this incredibly complex life and universe understandable to our level of comprehension so that we recognized that this immense universe is all necessary for this one, unique, tiny blue planet?, And wouldn't he have given us a holy book to verify his existence and detail his plan to get humans to the next life? Oh….wait a minute….That is exactly what he did.

Chapter 5

The Fab Five:
Abraham, Moses, David, Paul, and Melchizedek

Context and connections help construct a solid spiritual ideology.

In addition to the information already presented, it is time to construct an accountable spiritual ideology. The people through whom this spiritual ideology is established are actual historical figures considered to be significant by the Abrahamic faiths.

Abraham 2000 B.C.

The most interesting man in the world – or at least the most complete. Many think of Abraham as an old man, but in his prime he was a stud. He was extremely successful financially, militarily, and as a respected leader. Above all, he had integrity. Genesis chapter 14 describes how he took 318 of his trained men, born in his household, to defeat five evil kings who were terrorizing the region. As is customary, the victors had the right to the spoils. While he allowed his men to retrieve whatever they wanted belonging to those they had conquered, Abraham made it clear he would not take a single item. This man had incredible devotion and faith in God. He wanted everyone to know that his wealth came directly from God's blessing and nothing else.

Jews, Christians, and Muslims all point to Abraham as their patriarch. This man's man was the one through whom God established His covenant to redeem any individual and allow her or him into the afterlife. God promised a son to Abraham in his old age. After the boy reached a certain age, God asked Abraham to sacrifice the boy to Him. It was a brutal decision that had to be made. Abraham was faithful. He was lifting the knife in the air to complete the request when God told him to stop and, instead, to sacrifice the ram whose head was stuck in a nearby thicket [a sign of the Messiah ultimately having his head in a crown of thorns]. (Gen 22) God had found on Earth a man of character willing to sacrifice his son. Thus, God had the right to sacrifice His son for the sake of mankind. Two thousand years later, the seed from Abraham, his great-great-great ... grandson, Jesus, was sacrificed for mankind. And it was done in nearly or exactly the same location where Abraham was asked to make his sacrifice.

Moses 1500 B.C.

Moses was the one used to deliver God's people from slavery in Egypt. His life intentionally [on the part of God] parallels the life of Jesus. Both escaped death as infants. Moses' mother chose to put Moses in a basket and carefully send it in the river where and when Pharaoh's daughter could find him. Pharaoh's daughter did find him and raised him as her own. Thus, Moses was a Hebrew, of the enslaved nation, but was raised as a prince of Egypt. In his later years, Moses became the deliverer, promised to the Hebrews for centuries. He battled an enormously powerful Egyptian empire and the forces of evil behind it to free many thousands from slavery. (Gen 7 – 12)

Jesus, who also escaped death as an infant, would later take on the ultimate spiritual empire of evil and deliver into the afterlife anyone who accepts his act. Jeremiah prophesied a loss of children around the time of the Messiah, and the Gospel of Matthew records a decree by King Herod to have all young boys put to death in hopes of killing a newborn king born in the area. [Matt 2: 16 – 18] [Note: corroborating evidence for the slaughter of the innocents by Herod depicted in Chapter 2 of Matthew is difficult to find. One of the best sources for information around that time are the writings of Josephus, the Jewish historian of the day. Though he explains in his writings both the early success of Herod and his later murders of family and others, Josephus does not record the killing of boys under age two in this context. It does not mean that it did not happen. It simply means that it is for now unverified. Given the more dramatic executions of Herod's reign, it might have gone virtually unrecorded at the time. And such an action would lack historical significance until more than thirty years later when Jesus was crucified. Therefore, Josephus may not have been privy to such an event. And his mission was the recording of Jewish history and not the recording of the emergence of the Christian church.]

The Law was given to Moses, who recorded it among the first five books contained in the Tanakh. In these five books are key depictions of the creation of the universe that match what today's scientists suggest. In it are descriptions of events and people groups and locations that have helped archaeologists locate sites in recent times. In his message are the ten commandments and other rules that governed a society needing to endure the threats it would face. In these books is the Law, the ridiculously knit-picky set of rules that convinces us of our shortcomings morally. And within the Law and procedures practiced, were the justifications, revelations and prophecies of the plan - the blood payment from God Himself - that would usher humans into the afterlife.

David 1000 B.C.

David was an intriguing person of passion, commitment, bravery, and humility. His life and relationships are the makings of a very complete and compelling novel. In fact, if one claims the Old Testament of the Christian Bible [Tanakh] is boring, that person should read 1st and 2nd Kings and 1st and 2nd Samuel. It describes David's emergence as a chosen boy, his incredible bond with the king's son, and his years of devotion to a king-gone-mad. It describes his bravery and rugged life as a military figure and the years living as a fugitive. Eventually, David became king. But he later committed adultery and had that woman's incredibly dedicated husband put on the front lines of war so that he would be killed. David wore his emotions on his sleeve, took responsibility for his actions, and displayed a child-like devotion to God.

He wrote some of the more critical prophecies of the Messiah, including quoting Jesus on the cross 1000 years ahead of time in Psalm 22: *"My God, my God. Why have you forsaken me?"* The first use of crucifixion according to historians was 800 B.C. Thus, David shared details of crucifixion 200 years before crucifixion was even devised on planet Earth.[1] The recordings of Moses and David are given special significance in Islam as opposed to the way in which other books of the Tanakh are viewed.

Paul [Saul of Tarsus] 1st Century AD

The conversion of Saul of Tarsus – from a zealous Jewish Pharisee and Roman citizen who persecuted early Christians to one who gave up his wealth and stature to travel the world sharing why the Tanakh clearly points to Jesus as the Messiah – is very strong evidence for the resurrection of Jesus. Saul was a no-nonsense man. He was a Roman citizen and a highly educated Jew. He was devoted to God as he perceived God through the study of the Tanakh and via tradition as a Jewish Pharisee. He originally assumed that Jesus was a fraud. This powerful Saul of Tarsus was present and supportive of those who stoned to death Stephen, the first Christian martyr. Saul felt early Christians were heretics. He was rounding them up and throwing them in prison. Early followers of Christ were terrified of him. The Book of Acts in the Christian New Testament shares the account of Saul on the road to Damascus to round up more Christians. Jesus had already been crucified and many claimed he had risen from the dead.

"As he [Saul] neared Damascus on his journey, suddenly a light from heaven flashed around him. He fell to the ground and heard a voice say

to him,
'Saul, Saul, why do you persecute me?'
'Who are you Lord,' Saul asked.
"I am Jesus, whom you are persecuting.," he replied. 'Now get up and go
into the city and you will be told what you must do.'" [Acts 9: 3 - 6 NIV]

It took time for the reticent Christians to accept Saul as one of them when he first converted. But Saul's undeniable devotion confirmed his sincerity. He soon became known as the Apostle Paul. It appears part of God's plan to have an intellectual who had wrestled with his own bias and ideology become the most significant and foundational teacher in all of Christianity. It was Paul (formerly Saul of Tarsus) who let the Jewish believers know that scripture declared the Messiah to be a light to the Gentiles as well as to the nation of Israel. (Acts 13.47, Isaiah 49.6)

He claims to have met the risen Christ. He gave up his stature and wealth. He studied the Tanakh intently most of his life. His humility allowed him to question himself and articulate God's plan. And for years, mending tents to pay for his needs, he went throughout the Middle East, Europe, and Asia instructing all who would listen regarding God's plan. He fled for his life on many occasions and was arrested more than once. Ultimately, Paul was beheaded by the Romans for his faith in Jesus Christ. He is uniquely qualified to teach us all.

The first four of the Fab Five
The first four of these five figures were critically important and faithful. They had tremendous devotion toward God. Yet they all made huge mistakes. Abraham tried to help God's promise of a son in old age by producing a son through a maidservant. (Genesis 16) The two people groups – Arabs and Jews – have been at odds for most of the centuries since. Moses killed a man and fled. (Exod 2. 11 – 15) David committed adultery and plotted to kill the woman's husband. (2 Sam 11) Paul miscalculated who God was, and was responsible for the deaths of many

believers. The first four key players do not have clean records. This, too, is by design and further confirmation of how - even the best among us - fall short of God's standard. This awareness alone ought to instruct us to be less judgmental and more humble as believers and followers.

Melchizedek, the fifth and greatest of the Fab Five

Moses recorded in Genesis (Gen 14.17 – 24) the meeting between Abram (later to be called Abraham) and Melchizedek before the establishment of the covenant [promise] between God and man around 2000 B.C. Melchizedek was a high priest and king of Salem. He blessed Abram, Abram gave him a tithe, and then the two broke bread and shared wine. Recall how much devotion and integrity Abram displayed toward God. It is enlightening that Abram gave Melchizedek a tithe – presumably one tenth of his wealth. This Melchizedek is very significant. David claimed that the Messiah would be a priest in the order of Melchizedek (Psalm 110.4).

The writer to the Hebrews in the Christian New Testament joins David in declaring Jesus – the Messiah – as being a priest forever in the order of Melchizedek – not an earthly priest. In Hebrews Chapter 7, verse 3, the writer proceeds to give us the final insight into the identity of Melchizedek. He was *"without father or mother, without genealogy, without beginning of days or end of life, like the Son of God, he remains a priest forever."* This is not a created being. This is someone who has no beginning or end. This implies Melchizedek exists beyond our space-time dimension and is part of the godhead – the Son, Jesus. Melchizedek is Jesus.

Recall the idea that God is one God with three distinct persons in tandem. Moses recorded God stating He would *"make man in our image."* Moses also met this God who referred to Himself as I AM. In John's New Testament Gospel, he records Jesus nearly being stoned to death. Jesus said in response to a faithless comment, *"I tell you the truth, before Abraham was, I AM!"* Abraham broke bread and had wine with Melchizedek [Jesus] just prior to God establishing the covenant with man through the willingness of Abraham to sacrifice his son. Two thousand years later, Jesus [Melchizedek] broke bread and shared wine prior to the fulfillment of the covenant with mankind.

Abraham and Paul met Melchizedek before and after his human life, respectively. Moses wrote signs and prophecies about Melchizedek into the Law, and David recorded the death of Melchizedek - the Messiah, Jesus – 1000 years before it happened.

The Fab Five are connected via Melchizedek [Jesus] before and after his time as a man. The four men lived hundreds of years apart, yet there is no doubt of God's impact in their lives and the significant roles they play in outlining and pointing to God's plan and to God's grace... *given to us before time began.*

Chapter 6

Time and the Holy Book

How can God be in control but give humans a free will? And when God referred to himself as I AM, why was it a very concise and accurate description?

Paul gives us a scientifically accurate notion of our time in this universe. In 2 Timothy 1:9, he suggests that this grace in Jesus was given to us before time began. Scientifically and from a point of creation, our space-time dimension began with a singular, big bang. Paul is letting us know that God existed before our time began. In other words, God exists in and beyond our space-time dimension. It is impressive that someone in the first century A.D. would express such a notion.

Visualizing dimensions helps us better understand the implications of multiple dimensions of time. Below are three figures. The first is one-dimensional. It is a line. It has length, and that is all. The second is two dimensional. It is like a rectangle. It has length and width. We measure its area to denote its size - such as 24 square feet. The third is three dimensional. It has length, width and depth. It has surface area on all 6 faces, but its interior contains space or volume. It is measured in cubic units like a freezer that has 24 cubic feet of volume. Considering these as dimensions of time rather than space allows us to ponder God's relationship to us.

The scientific nature of our universe suggests activity in nine dimensions of space within the first split second of the universe.[1] Each dimension is perpendicular to the other eight dimensions.

Time line

Time Area

Time Space

Good luck visualizing that. But consider that a Creator exists beyond our space-time dimension. The implication is that this creator exists in at least one more dimension of space and at least one more dimension of time than we have in our created universe. Him existing in at least one more dimension of time helps to reconcile many questions we might have regarding spiritual matters. To my knowledge, no one has been able to explain how multiple dimensions of time might exist without introducing significant inconsistencies. However, the evidence of various predicted occurrences already discussed as well as some covered later in this book suggest input from a being not constrained by our one dimension of time.

Read the remainder of this chapter very carefully. It explains how God could have infinite time in any one moment of our time from which he can work. It also discusses how God exists all at once in our past, our present, and in our future. It helps us grasp how he has control while giving us a free will, because he knows every action anyone will ever make – good and bad. Finally, it helps us recognize spiritual beings can exist in another dimension of space; and yet, God ultimately knows every move they will ever make. It's like playing a game where God sees everyone's actions in the past, present, and future. He is ultimately in control of the game and is destined to win.

There are various passages in scripture and in the New Testament that speak of another dimension of time.

Paul suggests that God existed *before our time began*. In the letter to the Hebrews in the New Testament, the writer [unsure if it was Paul or someone else] claims that *"Jesus is the same yesterday, today and forever."* (Hebrews.13.8) Recall that this same author wrote about Melchizedek (Jesus), that he had *"no beginning of days and no end of days."* [Hebrews 7:3] These are all descriptions of God existing in at least one more dimension of time. Of additional interest is a prayer Jesus made to God, the Father recorded in John 17: 4, 5: *"I have brought you glory on earth by completing the work you gave me to do. And now, Father, glorify me in your presence with the glory I had with you before the world began."*

Consider the rectangle on the next page – like a sheet of paper – to be two-dimensional time. Let the line – like a string – represent the timeline of the universe. The string is sitting on the rectangle. The left end of the string is the beginning of our universe – the beginning of our time. The right end of the string is our current time extending into the future. The string will continue to increase slowly to the right as time proceeds. The rectangle gives an idea of what God can do if He exists in a second dimension of time. In this scenario, he is present anywhere on the rectangle simultaneously.

And from the perspective of our timeline, He is at all times at once – *the same yesterday, today and forever*. At any one moment of time along the timeline of our universe, God exists there. He can operate in the lives of 7 billion people simultaneously. And He simultaneously exists in the past and in the future. This is how God can impress upon David to write a Psalm describing Jesus on a cross 1000 years before it occurred. At this very moment in our lives, God can be here with us, back at the cross with

Jesus and back in 1000 B.C. with David. Nothing
surprises him, allowing Him to be in control while giving people a free

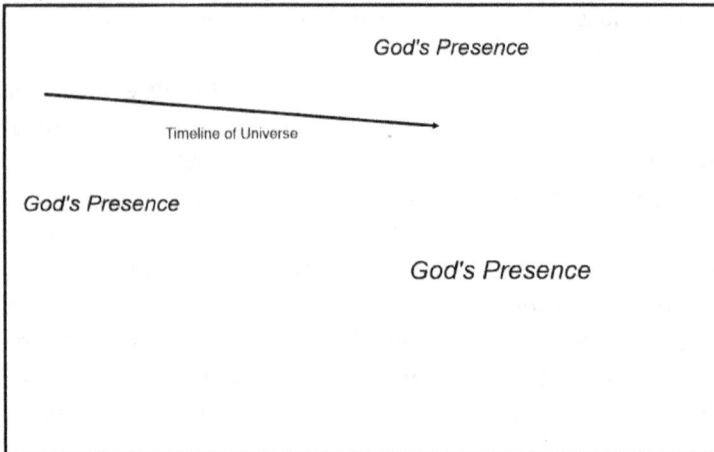

God's Presence

Timeline of Universe

God's Presence

God's Presence

will. This is likely why God described Himself to Moses very concisely as
I AM. God just is... everywhere... simultaneously. I AM.

Jesus is one of three persons of God and was nearly stoned to death by
religious leaders when he said, *"Before Abraham was born, I AM."* [John
8:58] In fact, looking at the few verses preceding his declaration, we are
brought full circle back to Jesus [Melchizedek] with Abraham 4100 years
ago. Here is the dialogue recorded between Jesus and some of the biased
religious leaders of his day:

Jesus:
*"'Your father Abraham rejoiced at the thought of seeing my day, he saw
it and was glad.'*
*'You are not yet fifty years old,' the Jews said to him, 'and you have seen
Abraham!'*
'I tell you the truth,' Jesus answered, 'before Abraham was born, I AM!'
At this, they picked up stones to stone him, but Jesus hid himself, slipping

away from them.'"

God being I AM is ultimately in control in our present universe. Nothing can separate us from Him – not even time. And Paul in his letter to the Romans stated, *"For I am convinced that neither death, nor life, neither angels nor demons, neither the present nor the future, nor any powers, neither height nor depth, nor anything in all creation, will be able to separate us from the love of God that is in Christ Jesus our Lord."* [Romans 8:38,39]

In Luke's Gospel of Jesus Chapter 24:44, he records the words of Jesus after he had risen from the dead.
"This is what I told you while I was still with you: Everything must be fulfilled that is written about me in the Law of Moses, the Prophets and the Psalms [Writings]." The Law, the Prophets and the Writings [Psalms] are the Jewish Tanakh.

There are statements in the Tanakh (Old Testament of The Bible), that we know are scientifically accurate today but should not have been known thousands of years ago, when they were recorded. These scriptural passages support the existence of a being from beyond our space-time dimension influencing the writers. For example, if an ancient writer was going to describe how God formed the universe, one would expect a description of how God *stretched* out the universe – past tense. The recognition of an expanding universe is a fairly recent scientific discovery. However, many passages in the Tanakh claim that God *stretches* out the universe – present tense. Today, we know this to be scientifically valid. Isaiah 40:22 is a very accurate depiction of our universe today. "He sits enthroned above the circle of the earth, … He *stretches* out the heavens like a canopy, and spreads them out like a tent to live in."

Such statements, in addition to clear descriptors of events that later took place in the life of the Messiah, were written to let future generations know that God influenced the writers of the Tanakh. As important as the Tanakh was to the hundreds of thousands of Jews before Christ, it is much more crucial to the billions of souls on the planet today. Other passages referencing the expansion of the universe include Psalm 104: 2, and Zechariah 12: 1. Job 9: 8, 9: *"He alone stretches out the heavens and treads on the waves of the sea. He is the maker of the Bear and Orion, the Pleiades and the constellations of the south."* Today we realize that these constellations move further apart as part of a universe that is not static – it is expanding. It indeed *stretches* out.

One other point worth mentioning from the 9th verse of Job 9. It says that God alone treads on the waves of the sea. Recall that Jesus was reported to have walked on water. (John 6: 16 – 24) This also identifies him as God.

The Genesis account describes a beginning of time – something which is unusual in holy books but scientifically accurate.[2] And if one reads the passage with literary integrity, it is in line with today's model of the evolution of our universe from 13.7 years ago until humans emerged.

Genesis 1:2 gives us the perspective from which to visualize the passage. *"The Spirit of God was hovering over the waters."* (Gen.1.2) Reading the passage from the perspective of being just above the surface of the water and the planet causes the scriptural account to line up with a scientific model of how the Earth evolved. The combination of verbs used in Chapter 1 of Genesis implies that the heavens and Earth were created brand new out of nothing seen – *"In the beginning, God **created** the heavens and the earth."* (Gen.1.1)[3] But a different verb combination is used to describe the emergence of the sun.[4] It is why translators of the original Hebrew chose the words, *"**Let there be** light,"* (Gen.1.3) as opposed to God *created* the light.

The Hebrew text translated *let there be light* is conveying that the light was already in existence but previously not seen.[5] This is because the Earth was initially covered by dense clouds like Venus – which the Book of Job correctly describes. (Job 38.9)   Early on there would be day and night, but no visible view of the sun or moon. It was during the Hadean Period of asteroid strikes that the dense cloud cover was broken up, and the sun became visible. Thus, the writer of Genesis records on day four – the fourth long period of time – that the sun was finally visible. *"**Let there be** lights in the expanse of the sky"* (Gen.1.14)

Once again the author chose not to say that the light was *created* at that point in time. Instead, the verbs imply that the sun and the moon and the stars existed already but now were literally being brought into view from the surface of planet Earth. The precise choice of verbs by the writer matches today's scientific model of cloud cover, the development of an atmosphere and water cycle as the cloud cover was broken up, and then the sun eventually becoming visible.[6] This is remarkable for a book written 3500 years ago.

Chapter 7

The Law: Misunderstood but Vital

The Law and what it helps us understand is challenging and revealing. But if not perceived accurately, it can become a stumbling block for individuals and cultures.

The Law given to Moses is contained in the Torah, the first five books of the Jewish Tanakh and Christian Old Testament. Ironically, the Law given to Moses in 1500 BC is foundational to our relationship with God and our acceptance into heaven; yet, it is often avoided or misrepresented in churches. The Tanakh [Old Testament] is referred to as The Law, The Writings, and The Prophets. Some people use it as a club to rule over others. Some use it to justify themselves. Others assume it is so outdated and harsh that it can be ignored. But the Law contains the plan to transform us, and it begins by letting us know how far short we fall when using it as the standard from which to judge ourselves.

Paul, the ultimate scholar, walks us through this very important portion of the Tanakh. It is wise to read the entire Letter to the Romans by Paul in the Christian New Testament. It is a dissertation on righteousness achieved through Jesus and the promise made to Abraham in connection to the Law. It is an explanation of how Jesus is the fulfillment of the law. Ultimately, it is faith in Jesus – discovered through humility and integrity – that ushers us into the next life.

In Romans Chapter 4, Paul reminds us that Abraham was not justified through the law. The law came hundreds of years later. Abraham was justified by faith. In Romans 4:15, Paul states that the *"... law brings wrath."* In Chapter 7, Paul shares a passage that reveals one of the main purposes of the law.

"What shall we say then? Is the law sin? Certainly not! Indeed I would not have known what sin was except through the law. For I would not have known what coveting really was if the law had not said, 'Do not covet.' But sin, seizing the opportunity afforded by the commandment, produced in me every kind of covetous desire. For apart from the law, sin is dead. Once I was alive apart from law; but when the commandment came, sin sprang to life and I died. I found that the very commandment that was intended to bring life, actually brought death. So then, the law is holy, and the commandment is holy, righteous, and good.
Did that which is good, then, become death to me? By no means! But in order that sin might be recognized as sin, it produced death in me through what was good, so that through the commandment sin might become utterly sinful." [Romans 7: 7 – 13]

Therefore, the Law helps us to recognize our shortcomings.

When initially revealed by Moses, the Law helped the Jews of Moses day to live a strict life allowing them to survive against the threats of other cultures infiltrating, conquering, or annihilating them. Ten of twelve tribes did not survive. In addition, the Law quantified the meaning and ritual of sacrifice, ultimately pointing to the sacrifice of God the Son, Jesus.

Recall how David in Psalm 22 referenced many details of the crucifixion. *"A band of evil men has encircled me, they have pierced my hands and my feet. I can count my bones; people stare and gloat over me. They divide my garments and cast lots for my clothing."* Additionally, when David mentioned that his hands and feet were pierced, he was referencing a form of capital punishment – crucifixion - that would not be devised on Earth for another 200 years.[1]

God let the other four of the Fab Five know that the Son [Melchizedek,

Jesus] was to be crucified. In 2000 B.C. as Abraham and his young son were walking toward the place where the sacrifices were made, his son asked him, *"'The fire and the wood are here, but where is the lamb for the burnt offering?'*
Abraham answered, 'God himself will provide the lamb.'" (Gen.22.8)

With Moses in Egypt in 1500 B.C., God instructed the Hebrews to take blood from the Passover lamb and put it on the sides and tops of their doorframes to prevent death. (Exod.12. 7) This was a sign of the true Passover lamb, Jesus, who would have his blood put on the wood of the cross to take away the sins of mankind and create the portal to the afterlife. [How many of us realized that the Jews were crossing themselves long before the Catholics took up the gesture?]

According to the law, the Passover lamb selected was to be without blemish; and none of its bones were to be broken. This is why David, in 1000 B.C. when describing the future crucifixion of the Passover lamb, mentioned in Psalm 22 that *"I can count my bones."* It is also why John, in Chapter 19 of his gospel, recorded that the soldiers needed to get the three men down from their crosses, because the special Sabbath of the Passover celebration would begin at sundown. Breaking the legs of those on the cross prevented them from pushing up to get air. Once the legs were broken, the lungs would fill with water, and death would occur.

John recorded how they broke the legs of the two thieves, but that Jesus was already dead; therefore, they did not break his legs. This is the fulfillment of prophecy recorded by both Moses and David. Additionally, John recorded that they pierced the side of Jesus to confirm that his lungs had filled with water and that he was dead. This was also a fulfillment of the prophecy of Zechariah 12:10.

Returning to the law in Exodus Chapter 12, Moses instructed that in future honor and celebration of the Passover in Egypt, the Passover lamb was to be selected on the 10th day of Passover, sacrificed at twilight on the 14th day, and the 15th day was to be a special Sabbath celebration. Recall what John recorded regarding the death of Jesus. They had to get the three off the crosses, because the next day was the special sabbath – day 15. Day was ending. Twilight had begun. It was the 14th day of Passover when Jesus was crucified [sacrificed] at twilight. The next day was the special Sabbath, the 15th day of Passover. Additionally, on the 10th day of Passover, Jesus rode into Jerusalem on a donkey [prophesied by Zechariah in Zechariah 9:9]. All of this was a perfect fulfillment of prophecies from 2000 B.C. through 150 B.C.

The way the events played out in the final week of the human life of Jesus is mathematically profound. For 1500 years, Jewish culture practiced for the ultimate entry of Jesus into Jerusalem. Moses instructed the Israelites that each year would begin with Passover. On the tenth day, each man was to select a Passover lamb. According to John's gospel, Jesus was near Jerusalem in Bethany six days before the Passover feast, which occurs on the 15th day. (John 12: 1) In John 12: 12, 13 the account continues. *"The next day [this would be the 10th day of Passover] the great crowd that had come for the Feast heard that Jesus was on his way to Jerusalem. They took palm branches and went out to meet him, shouting, 'Hosanna [God save us]! Blessed is he who comes in the name of the Lord!"* Some historians have questioned whether it would have been palm branches that were held. The type of tree is irrelevant. Some have suggested that Jewish tradition had the high priest going out of the city to select the Passover lamb, and upon his return, the congregation would wave palm branches and shout *"Hosanna!"*

After the death and resurrection of Jesus, John and the disciples read accounts in the Tanakh of what the disciples witnessed regarding the Messiah's last days. In addition to seeing Jesus alive again, the most significant holy book gave many vivid details of what they witnessed. The disciples were reading a story that they watched play out, knowing that the story was written hundreds of years earlier.

The impressive prophecy regarding the entrance into Jerusalem is found in Psalm 118. In order to grasp a greater understanding, one can reference another portion of the Law involving cities of refuge. Numbers 35 describes how the Law established six cities of refuge. If a man accidentally killed another man, he had to outrun his pursuers to a city of refuge, grab the horns of the altar, and plead his case. If it was determined that the death was accidental and not premeditated murder, the man had to remain in the city until a very significant event occurred. The event was the death of the high priest. Once the high priest died, the man was forgiven and free. Obviously, this is a sign of the death of the Messiah [Jesus, Melchizedek – the high priest].

Of further significance are the horns of the altar, where sacrifices were performed. Combining implications recorded in Exodus 21: 12 – 14 and 1 Kings 1: 50 – 52, men ran to the altar and grabbed the horns of the altar to plead their case. It is where they would either be condemned or freed by the death of the high priest.

Psalm 118 describes the entrance of the Messiah into Jerusalem. First, Psalm 118: 22 states, *"The stone that the builders rejected has become the corner stone."* The Messiah was to be rejected by the leaders of his day, yet he would become the foundation of God's kingdom for eternity. From there the Psalm declares what an important time it is; and it mentions that people would be yelling *"Save us!"* and *"Blessed is he who comes in the name of the Lord."* Psalm 118: 27 describes the scene 1000 years after the Psalm was written. *"With boughs in hand, join in the festal procession up to the horns of the altar."*

Referencing the horns of the altar ties the event to the death of the high priest in the cities of refuge. The prophecy did not specify that people would wave palm branches. It merely stated that the boughs of some tree would be in the hands of those escorting the Messiah into the city in celebration and preparation for the feast. Again, here is what the disciple John recorded:

"The next day [this would be the 10th day of Passover] the great crowd that had come for the Feast heard that Jesus was on his way to Jerusalem. They took palm branches and went out to meet him, shouting, 'Hosanna! [God save us] Blessed is he who comes in the name of the Lord!" That day, neither the crowd nor the disciples realized what would happen four days later. Four days later would be the completion of the promise to Abraham. Four days later, God's Passover lamb – Jesus - would be sacrificed. Six days later, in heaven – where the actual tabernacle and altar not made by human hands resides (Exod 25: 9, 40; Heb 8: 1 - 6), the high priest Melchizedek – Jesus – would sprinkle the blood of God on heaven's mercy seat to open the portal to the afterlife for any man or woman whose heart chooses to go there. Following this monumental occasion, the risen King – Jesus – would return to Earth to reveal himself to his followers and later to Saul of Tarsus. All these events began with the procession into the city on the 10th day of Passover.

Isaiah wrote a lengthy description of the life, death and resurrection of the Messiah in Isaiah 53 (circa 700 B.C.).
"He was oppressed and afflicted, yet he opened not his mouth; he was led like a lamb to the slaughter, and as a sheep before her shearers is silent, so he did not open his mouth.

By oppression and judgment he was taken away. And who can speak of his descendants? [He had no earthly children.] *For he was cut off from the land of the living; for the transgression of my people he was stricken. He was assigned a grave with the wicked* [he was crucified with two thieves], *and with the rich in his death* (he was buried in a rich man's tomb), *though he had done no violence, nor was any deceit in his mouth.*

Yet it was the Lord's will to crush him and cause him to suffer, and though the Lord makes his life a guilt offering, he will see his offspring [spiritually] *and prolong his days* [he will rise from the dead], *and the will of the Lord will prosper in his hand. After the suffering of his soul, he will see the light of life and be satisfied* [he will return to his spiritual form]; *by his knowledge my righteous servant will justify many, and he will bear their iniquities.*

Therefore, I will give him a portion with the great, and he will divide the spoils with the strong, because he poured out his life unto death, and was numbered with the transgressors. For he bore the sin of many and made intercession for the transgressors."

The Apostle John recorded the following after the resurrection in John 12: 16: *"At first his disciples did not understand all this. Only after Jesus was glorified did they realize that these things had been written about him and that they had done these things to him."* These were fisherman and lay people. They were not educated like Saul of Tarsus. They must have felt great faith and amazement as they later read the various prophecies in the Tanakh. They lived what they were reading.

There is another interesting detail from Leviticus 19: 5 – 9. The practice of the burnt Passover offering was to sacrifice it one day and eat it on that day or the next. But Moses warned that anyone who ate it on the third day was desecrating what is holy. This is likely because Jesus, the ultimate sacrifice, would rise from the dead on the third day.

Paul explains how God precisely established that Jesus [Melchizedek] could bear the transgressions of us all, and thus establish the portal through which we are able to accept and thrive in an afterlife. God included a specific provision in the Law recorded in Deuteronomy 21: 22 and 23. *"If a man guilty of a capital offense is put to death and his body is hung on a tree, you must not leave his body on a tree overnight. Be sure to bury him that same day, because anyone who is hung on a tree* [other interpretations of the Hebrew describe a pole] *is under God's curse."*

This is eventually how God was able to curse Jesus on a cross with the sins of mankind - and why Jesus would then utter the words, *"My God, my God. Why have you forsaken me?"* Paul beautifully ties it all together; and he puts the law into perspective at the same time. From his letter to the Galatians, Chapter 3:

"You foolish Galatians! Who has bewitched you? Before your very eyes Jesus Christ was clearly portrayed as crucified. I would like to learn just one thing from you: Did you receive the Spirit [the Holy Spirit, the third person of the Godhead] by observing the law or by believing what you heard? Are you so foolish? After beginning with the Spirit, are you now trying to attain your goal by human effort? Have you suffered so much for nothing – if it really was for nothing? Does God give you his Spirit and work miracles among you because you observe the law, or because you believe what you heard?

Consider Abraham: 'He believed God, and it was credited to Him as righteousness.' Understand, then, that those who believe are children of Abraham. The Scripture foresaw that God would justify the Gentiles by faith and announced the gospel in advance to Abraham: All nations will be blessed through you. So those who have faith, are blessed along with Abraham, the man of faith.

All who rely on observing the law are under a curse, for it is written: 'Cursed is everyone who does not continue to do everything written in the Book of the Law.' Clearly no one is justified before God by the law, because, 'The righteous will live by faith.' The law is not based on faith; on the contrary, 'The man who does these things, will live by them.' Christ redeemed us from the curse of the law, by becoming a curse for us, for it is written, 'Cursed is anyone who is hung on a tree.' He redeemed us in order that the blessing given to Abraham might come to the Gentiles through Christ Jesus, so that by faith we might receive the promise of the Spirit.

Brothers, let me make an example from everyday life. Just as no one can set aside or add to a human covenant that has been duly established, so it is in this case. The promises were spoken to Abraham and to his seed. The Scripture [Tanakh] does not say 'and to his seeds', meaning many people, but 'and to your seed,' meaning one person, who is Christ. What I mean is this: The Law introduced 430 years later, does not set aside the covenant previously established by God and thus do away with the promise. For if the inheritance depends on the law, then it no longer depends on a promise; but God in his grace gave it to Abraham through a promise.

What then was the purpose of the law? It was added because of transgression until the Seed to whom the promise referred had come. The law was put into effect through angels by a mediator. A mediator, however, does not represent just one party; but God is one.

Is the law, therefore, opposed to the promises of God? Absolutely not! For if a law had been given that could impart life, then righteousness would certainly have come from the law. But the scripture declares that the whole world is a prisoner of sin, so that what was promised, being given through faith in Jesus Christ, might be given to those who believe.

Before this faith came, we were held prisoners by the law, locked up until faith should be revealed. **So the law was put in charge to lead us to Christ that we might be justified by faith.** Now that faith has come, we are no longer under the supervision of the law.

You are all sons of God through faith in Christ Jesus, for all of you who were baptized into Christ, have clothed yourselves with Christ. There is neither Jew nor Greek, slave nor free, male nor female, for you are all one in Christ Jesus. If you belong to Christ, then you are Abraham's seed, and *heirs according to the promise."*

Paul was one of the first intellectuals to challenge his bias, confront his ideology, hold himself accountable to Scripture [the Tanakh] and to reality, and uncover the ultimate mystery and plan created by God. Today, more than ever, we need intellectuals who recognize what Paul is sharing to confront and change some damning group-thinks within the religious community as well as within the scientific community.

There is one path through the portal. It is precise, like God's universe and this life is precise. And yet, God in his grace, mercy and practical love, will allow anyone into heaven that has not heard of this Jesus and of this plan provided they follow the leading of their heart – the law written on their heart as Paul suggests at the end of Romans Chapter 2. But even they will make it to the afterlife only through the portal God established: the sacrifice of Jesus, God the Son. Therefore, Jesus is indeed, *"The way, the truth, and the life."* No one will come to the Father except through him and his act.

The law of Moses is often misunderstood. To get too hung up on the rules in the Law is to misunderstand its purpose and the manner in which we are to live - namely that we are to live by faith and the love and works displayed and produced from that faith.

Some atheists have thought that they logically destroyed Judeo-Christian ideas by "exposing" crazy ideas in the law. For example, some have referenced children being put to death if they are drunkards and gluttons. (Deut 21.18 – 21) Some atheists think Christians and Jews are crazy to follow a God who would command people to kill their children. However, such an argument ignores context. Ego, bias, and self-centeredness [EBS] can make intelligent people reach erroneous conclusions.

A simple understanding of the culture and age of accountability in the time of Moses helps us gain a more educated understanding of the situation. Unlike our nuclear families of today's industrial and technological age, in Moses day it was a patriarchal, agricultural society. Grandpa was in charge. If his 24 year-old son was defying what it took to preserve the tribe and was a drunk and glutton running perhaps with the women of the enemy and not at all willing to change, he was to be put to death. Moses was not talking about putting six-year-olds to death. He was establishing strict laws to preserve the Jewish race during a more tenuous time.

Chapter 8

The Not-so-Fab Five:
Pain, Suffering, Evil, Guilt, and Fear

This chapter is primarily for the large portion of the population who struggle with or are affected by these five forces.

My goal for more than thirty years has been to develop a spiritual ideology accountable to Science, Scripture, and History. And a huge question to answer is *why so much pain and suffering.* If Jesus is the Messiah and his teachings and Paul's teachings confirmed through the Tanakh are accurate, then we must accept the reality that God created a universe and life including pain and suffering and evil. He must exist in at least one more dimension of time and space and thus would know the horrific things men would do to one another. He must have known asteroid strikes would kill, tornadoes and earthquakes would kill, and men would kill. Blaming these things on fallen man equates to denying God's control and existence beyond our space-time dimension. In other words, if God was not aware these things would occur, then the God of the Tanakh and Bible does not exist. We may want to devise our own paradigm and believe God was unaware what would happen, but such a paradigm would lack accountability to Science, Scripture, and History. This may be the most difficult realization to wrap our minds around. But God is either who He claims to be, or he is not.

Pain and suffering are delicate subjects to address. How can God sit back and allow some of the things that are done to people? It would make it seem as though God is unloving or does not exist. It may help a little that he was willing to be born a human – a poor human. And it may help to know that he was willing to be severely beaten and then brutally killed. But if he is God, He had the choice to make things so different. So why didn't He? The answer only makes sense when we compare this short life to an afterlife that will be rich in relationships, creativity, and fulfillment - and will be free of an evil presence because of what is purged via this first life.

Those of us who struggle well through this life and come to the realization of what is available to us – through no great act we have committed but rather through God subjecting himself to extreme pain and bloodshed on our behalf – will enter the afterlife with a perspective that makes the afterlife greater than we could have imagined. Being free from those individuals on Earth who chose arrogance, selfishness, and exploitation will make possible a relational richness with others that will blow us away.

Compared to billions of years and more in an afterlife that just is, the life we live here is extremely short. The pain of this short life may be compared to the brief pain of receiving a shot of medicine that provides years of good health. And the many joys of this life give us a taste of the life to come.

Some people have had unspeakable things done to them. God did not create those individuals to be targeted, but he certainly did allow it. And it is not scripturally accurate to say, "Everything happens for a reason." Yes, generally pain and suffering must be part of this life for various reasons. But individually, this is an inaccurate and cruel statement to share with someone going through a crisis. The teachings of Paul [Romans 8:28 in particular] suggest that *"In all things, God works for the good of those who love him."* And Jesus said, *"In this life you will have tribulations. But be of good cheer. I have overcome the world."* (John 16.33)

When life gets difficult, some of the most valuable people available are those who went through a crisis successfully. Often those with beautiful hearts were once subjected to things we would not want to endure. God may use their pain to increase empathy and develop keen understanding.

Pain and trauma can be debilitating, but it can also produce some of the most desirable human traits and interactions. In a strange but intentional way, people on Earth who go through a tragedy or trauma with another they trust, experience a depth of relationship that others may not. Many world relief workers make this observation. They have witnessed immense valor, courage, and goodness from persons who have been devastated, but who selflessly attended to the needs of others.

Consider also military personnel who survive trauma and challenges with others in the armed services and build a bond that otherwise would not exist. Unfortunately, the flip side is that they may have to endure lingering pain and trauma that few of us will experience. One thing expressed by active-duty military men and women returning from conflict is their frustration with those who take so much for granted. It is an important observation. When our lives are too easy, we become lax in many respects. This may be why it is easier for those in poverty to pursue a spiritual journey compared with some who are wealthy.

A note to those in the military who suffer from the effects of what they have witnessed: *such suffering does not make you weak. You have experienced things that the rest of us pray we never do. Sometimes the platitudes or ideas we had implanted in us growing up do not match the reality of life or of war. In such cases, abandon those platitudes and ideas, because God is the God of reality. We can question Him at times for* *creating a world with the horrors of war; but the horrors of war are part of reality. It is okay to rant at God for what you experienced and for the dark aspects of this life. There exists a strange faith in doing so, and it can be our honest emotion at times.*

You should not be forced into a paradigm that ignores what you endure. Pain does not permanently leave us in this life. In a moment's notice, we can feel grief that is nearly unbearable. The only things that make sense IF there is a God in control and given the brutality of this life, is that God knew it would happen and did nothing to stop it. So question, rant, and grieve...until it starts to sink in that – in light of an afterlife – this life is precisely what is necessary to optimize the next life. This life – that is sometimes heaven-on-Earth and sometimes hell-on-Earth – must work its purpose to pave the way for an afterlife of billions of free-willed beings that can flourish in a far more powerful realm orchestrated by love and trust.

You are not at all weak for the pain you endure on behalf of the rest of us who reap the rewards of your sacrifice. My prayer for you is that you recognize your humanness – the good and the bad within us all – and cling to your virtue and character even when you may think it has left you. Hang on during the dark moments and be determined to move forward. Get help and encouragement from others. You are not weak in doing so. You are wise. Do it for you, for your loved ones, and for the other members of the military who need your example and encouragement. It is truly a struggle and sometimes not at all pretty. And yet, if we struggle well, it can be a beautiful part of our life. May God watch over you. Thank you so much for what you have accomplished for so many. And may God work through wise, effective individuals and organizations to better care for military personnel harmed in their service.

Too often it seems our military personnel are injured twice. Once during military action and again if government agencies turn their backs and fail to take responsibility for some health problems caused by military decisions or practices.

God created the universe exactly the way He intended. He made sure that nature would point to him, whether some can break away from their bias and ideology long enough to consider it. He provided a holy book that is accepted still today by half of the world's population. He provided signs and prophecies through writings and actual life events on the part of many key individuals. And he himself is the person 2000 years ago that so many still point to today as the embodiment of love. The plan is clear and substantiated for those who make an earnest attempt to find it. Ultimately, it is our decision. Will we choose to accept the incredible gift from God that cannot be earned, or will we choose an ideology that fails to acknowledge the purpose of this short life?

God wants the greatest number of people to choose the gift. It is this desire that necessitates the proper balance of evil, pain and suffering here. Mathematically, it provides for the greatest number of souls to accept the gift. A lack of evil, pain and suffering (that might otherwise catch our attention) would be evidence of a God who does not care enough to get the greatest number of people to choose correctly.

Unfortunately, the free will, insecurity, self-centeredness and ego we humans possess can unleash horror on others and can subtly form a perspective on the part of its recipients that can haunt and harm them throughout life. Perhaps nowhere is this more true than the impact a parent can have on a child. A parent can desire to do everything right and want the best possible life for his or her child. But parents live with their own stresses and past hurts as they inevitably cause harm in various circumstances throughout the life of their children. Even the best parents have regrets. But they also have a way of training children to recognize that bad moments will occur. Forgiveness and understanding in these experiences can help us all to interact better and to know how to take proper responsibility for our actions.

But in cases where parents have become self-absorbed and inhumane, their abuse of their vulnerable children will create a living hell those children may struggle with into adulthood. When a parent – who is the natural protector – is the very one to betray a child, the difficulty for that child at any age in life to put trust in others to help them through the hurt is a second assault on their livelihood.

For the many who live with deep hurt from such circumstances, please consider this. The way out is always through a perspective of wisdom and understanding. It is the mind and the heart that ultimately must come together. It is a war that must be waged against the inaccuracies of what our minds or emotions may tell us. Unfortunately, or fortunately, there are many who – though the experiences are always unique – can relate to the injustice and to the battle that must be waged. Not every question has an available answer. But enough questions can be answered; and we must build our perspectives on reliable information to climb slowly out of the darkness and hurt that sometimes feels beyond bearable. Seek the help of others. Love yourself unconditionally and hang on to hope.

A study of the promises in the Tanakh and the recordings of things Jesus said and things he did help us to recognize that Jesus is part human and part God. And he provides a tangible vision of who God is. Though many justifiably are frustrated with certain Christian behavior today and over the centuries, few would question the motives, wisdom, justice and goodness of Jesus. He destroyed religiosity and connected with people in the manner they needed. He hung out with those that others rejected. And he had a way – a godly way – of saying or doing the ideal thing that helped open eyes to reality. He ultimately conveyed the wisdom and the understanding that powerfully put things into perspective and permanently changed lives for the better: for eternity.

When our perspective is in line with reality, then we can move forward in whatever circumstance and time we have left to live in a way that makes sense. If we are wise and undaunted, our allegiance then becomes to Jesus and to His ways. It is not necessarily to the religious ways that all too often deviate from the example and teaching of Christ. It is upon Jesus and the God that understands pain (and allowed it) that we cast ourselves. We can yell at Him, question Him, and wrestle with Him. He can comfort us and give us hope. And our pain can be used to help others.

And yet, the pain is never truly gone. It can return in an instant. But if we are willing, it has its place, and it can produce a positive power that other experiences cannot. But it can take time – two steps forward and one step back. Life can be brutal. But it is also beautiful.

Before he was captured, Jesus knew the horrible things the Romans would do to his body. He felt it.
"And being in anguish, he prayed more earnestly, and his sweat was like drops of blood falling to the ground." [Luke 22: 44]

The good in many of us wants to suggest that God could never allow anyone to be hurt. Perhaps we want to believe there is only heaven and no hell – whatever hell is. Some teach that, once we accept Christ and follow him, our lives will be without harm the rest of our days. But none of this lines up with what we observe in life or witness globally.

When people look at this life without the context of an afterlife, it makes no sense that a loving God would allow such hardship. There would be little reason to serve God if we lived this life, died, and there was nothing more. Such a life would point to a cruel God. However, in light of an afterlife and with a recognition of what this challenging life accomplishes in preparing the sort of existence yet to come through the suffering and experience of this first life, the Plan established in the Tanakh points to the work of a highly calculated, loving creator.

We may not think we like this world and this life. In fact, none of us always likes it. But this is reality. And a loving God, who experienced first-hand a greater brutality than most of us will ever face, is in control. The chaos and hurt in the world today is not evidence of no God or of a God not in control. On the contrary, it is what He made and is exactly what is necessary. He tells us how it is. He let us know that we will have tribulation. He came down and experienced it. He promises us an absolutely beautiful afterlife.

When some ponder their mortality, another nudge is provided. The physician Luke recorded these words from Jesus: *"I tell you, my friends, do not be afraid of those who kill the body and after that can do no more. But I will show you whom you should fear: Fear him who, after the killing of the body, has power to throw you into hell."* [Luke. 12. 4, 5] Hell is a reality taught by Jesus. If true, it may be assuring to know that tormentors on Earth will receive a much more horrific torment than they ever inflicted in this life. But there is yet the recognition that God allows tormentors to exist.

Consider that on average each day 16,000 children age five and under die of malnutrition, famine, lack of clean water, disease or conflict.[1] Thankfully, this is an improvement since 1990, when 35,000 children died per day on average.[2] Perhaps part of why God allows the deaths of these children is to make a very relevant point. Just as He sits back and allows these children to die in this life, He will sit back and watch people die the second death in hell. We collectively choose to let kids die with lack of sufficient action, and God sits back and allows it. The good news is that these young kids who die – having not yet reached an age of accountability – will join God in the afterlife. The bad news is that, according to the teachings of Jesus and the implication of the Tanakh and the need for God's plan, there is a hell to be experienced by those who avoid the humility and diligence required to recognize, ponder, and accept the purpose and plan of this life.

Prove that there is a hell? We can't. Prove that there is not a hell? We can't. But we can prove beyond a doubt that a God exists and that the plan involving Jesus is the designated passage to the afterlife. And the implications along with the teachings of Jesus and Paul suggest both a heaven and a hell.

This life, with its joys and blessings, is also a battlefield for souls. Living with understanding, integrity, purpose, love, courage and humility leads to the greatest good on this battlefield. The epiphany of God's plan and the humility to grasp it and communicate it is necessary for impacting others positively in the spiritual realm. It is why it is so sad to see the lack of intellectual arguments for the existence of God in tandem with the recognizable plan that paves the way to the afterlife.

Many educated individuals look at the simple faith exhibited by persons in impoverished countries and feel sorry for lives lived with such superstition. But the simple journey of some impoverished persons may lead to an adequate understanding and acceptance of the plan that many intellectuals fail to examine. Ironically, in the afterlife it will be many of those simple people who will feel incredibly sorry for the intellectuals who chose to ignore the evidence and who failed to question reality with the necessary diligence.

In 2Peter 3: 8 - 9 Peter states, *"With the Lord a day is like a thousand years and a thousand years like a day.* [Again, an indication of how God exists outside of our dimension of time.] *The Lord is not slow in keeping his promise, as some understand slowness. He is patient with you, not wanting anyone to perish, but everyone to come to repentance."*

85

Putting guilt and fear in perspective:

We need to fear God if we have not made a commitment to accept His plan. Once we have made such a commitment, the grace paid for by God himself covers us. We should not get mired in guilt if our hearts are moving in a loving, committed direction. John, the disciple of Jesus, clarifies this point in I John 4: 18, 19:

"There is no fear in love. But perfect love drives out fear, because fear has to do with punishment. The one who fears is not made perfect in love. We love because he first loved us."

Therefore, with the fear in life, he gets our attention; and then his perfect prophesied plan gives us the grace and assurance that fear now has a much lesser role in our lives moving forward. This brutal life and God's design of it and awareness of it should help us to put guilt in its proper place as well. Guilt is debilitating. It prevents us from walking in better understanding and in love with God. It turns the focus too much on ourselves as opposed to God, his provisions, and his desire to work through us for the sake of others in the time we have left. Unfortunately, in many religious communities and churches, guilt is heaped upon people in unhealthy ways.

Guilt is one of the first things people turn to when they do not comprehend God's reality. Many walk away from church because of guilt. Again, guilt puts too much emphasis on us as individuals. We should keep our eyes on the love of God and be led by love rather than by fear or by guilt. Remember the words of John: *"fear has to do with punishment."*

Obviously, there is a balance to be found here. Our conscience and the leading of the Holy Spirit can convict us when we are in error – which can happen quite frequently. Our response to such nudging can take us immediately back to a healthy, loving relationship with God. To obey [these convictions] is better than sacrifice. The way to progress is typically via a committed, loving desire to please God and renew our thinking to be more effective and to gain the wisdom that assures our minds and hearts that we are moving in the right direction relationally and practically. Guilt, promoted by many religious teachings, has sent people into mental wards and led some to take their own life.

For those who feel stuck in a pattern of guilt, here is a strange but helpful perspective:
God does not sin. But he purposefully created the beings that commit the most horrible atrocities. Therefore, why would one be consumed by guilt when God created people He knew would commit the most vile acts? Otherwise, God is not in control. Thus, God is somewhat on the hook here. This life is quite messy. At times it is a hell on Earth. You should not attempt to hide your sins from God. Instead, take them to Him. He knows exactly the problems commonly faced by people.

You have not freaked God out. Don't listen to well-intentioned Christians who say things like God and the Holy Spirit are so gentle that they cannot handle sin and must look away. Those are the kinds of ideas that build with other ignorant ideas to form a tidal wave of guilt that pummels people today. When you confess your sin, and through a spiritual epiphany commit yourself to God through the act of Jesus, God forgives your sin. You become a new creation moved by love and compassion and no longer are to be ruled by fear and guilt. Be moved to tears that God forgives, but do not stay in the mire of guilt. It does no one any good; and it is an unbalanced view of God's universe and his love for each of us.

Through Jesus, you can be "born again." It does not make you perfect in this life. It makes you forgiven. It gives you mercy. It replaces fear with love. We should renew our minds to this reality daily, because we have work to do. It starts with us allowing God to get our thinking in line with reality.

In bringing this chapter of heavy ideas to a close, an extremely important point needs to be made. The point is this: have some fun. We are to seek a balance of recreation in our lives. When it is time to take a break from work and serving others, take a break. Have a blast. Enjoy your friends and family. Yes, we need to build our spiritual understanding, serve when we are supposed to serve, support ministries and charities, and be there for others. But we are also to take time regularly to relax, enjoy, treat ourselves, our families, and our friends. The sky is more blue and food tastes better when we move through life in a purposeful manner with God and when we take time to appreciate.

Don't let fear or the need to prove something or a strange sense of inferiority or obligation incline you to over-commit. People with high empathy that are easily made aware of regrets in their lives are susceptible to over-doing it and to denying themselves too often and too much. If this describes you, reach out to others for perspective. You are a more effective and powerful you when you do what should be done and not more. And you are more powerful and effective when you can recreate and take time to indulge yourself occasionally.

Celebration, good food, and relationships are a prescribed and important part of your life. Just because the afterlife is the ultimate goal and reward, does not mean that the beautiful things God created here are not to be enjoyed. Balance is important. We should care for ourselves and care about others. Giving and living with purpose are keys to a happy, healthy life as are the moments we indulge ourselves and loved ones.

Chapter 9

200,000 Years of Humans

Humans differ from hominins, and a localized flood is possible both scripturally and scientifically while a global flood is not.

There are things we know and things we don't know regarding the spread of humans across the planet over the past 200,000 years. While archaeologists continue to uncover clues, a fanciful story is prevalent in churches. It is a story of a *global* flood involving Noah, who was instructed to take two of every animal aboard the ark. With billions of species at the time, this would have been quite impossible. Fortunately, this widely accepted story is <u>not</u> what the Genesis account conveys. The account in the Tanakh (Scripture) is of a *localized* flood – something that can easily line up with history and science.

When intently reading the account of the flood in Genesis, one should quickly notice in Chapter 7 verse 2 and 3 that God did not instruct Noah to take two of every animal. The ark was big, but not that big. Noah was instructed to take seven pairs of certain animals, two pairs of others and seven pairs of birds – and by implication, none of most animals. The flood was not global. He did not need to save all the world's species. Life for animals in other locations of the world was likely minimally affected. Noah and his family would be starting from scratch and would require certain animals.

Applying literary and logical integrity to the book of Genesis makes it easy to determine that a localized flood is being described. In the poetic Psalm 104, the promise is made by God that, once the waters gave way to the emerging land mass on Earth billions of years ago, *"never again will they [the waters] cover the earth."* (Psalm.104.9) Thus, a global flood in Noah's day would contradict this promise made elsewhere in scripture. In other words, the flood was not global. If it were, it would be a scriptural contradiction.

Contextual clues in the Genesis account also confirm the flood was not global. The Hebrew word *erets [Strong's H776] is* translated into *earth* in two key verses of the flood account. While it could imply the entire world, it more typically implies an area of land or ground. The phrase that God will *"wipe from the face of the **earth** every living creature I made,"* (Gen. 7.4) is later clarified in Genesis 8:9. *"But the dove could find no place to set its feet because there was water over all the surface of the **earth**."* (Gen. 8.9) When the dove went out, the hills were already exposed. Therefore, the surface of the earth could not be the entire planet. Thus, the passage interprets itself.

The face of the earth and surface of the earth were only a region of the planet. Geology and archaeology would not contradict a massive flood on the Mesopotamian Plain sometime in the past 80,000 years.[1] Determining when such a flood occurred is nearly impossible. A scant amount of evidence exists for such a flood in the Mesopotamian Region, because after thousands of years, such geological evidence is difficult to detect.

There were hominins [human-like primates] going back millions of years. In 2015, Prof Brian Villmoare of UNLV reported that a recently discovered 2.8 million year-old specimen of a hominin in Ethiopia was closely related to Lucy from 3.2 million years ago.[2] The BBC reported it as the "first human" discovered in Ethiopia. However, this is a misnomer. There are many species of the genus homo, but humans differ in many significant ways from Lucy and hominins. Most scientists still consider that we modern day humans emerged within the last 200,000 years.

Evidence suggests a large group of humans were killed in 70,000 B.C. in Sumatra, Indonesia when a colossal volcano erupted at Toba.[3] And one source claims evidence the human population on Earth dropped to as low as 40 pairs of males and females around this time.[4] Might the confidence interval be off just a bit, allowing for the notion that there were only 4 pairs left: Noah, his three sons, and their wives? Or did God only need to start over with his chosen line of people at that point in history? Were their humans that evolved in addition to God's chosen line through Adam? It is encouraging that today we can determine a bottleneck of humans [a very few humans left on earth] around one plausible time of a key localized flood.

The account of the Tower of Babel in the Tanakh is also not what many have been taught. First, it reads more like an allegory or legend. It is strangely nestled in between a much more academic flow of thought regarding the descendants of Noah. (Gen.11.1 – 9) The account of the flood incorporates very specific names and details. No specific persons are mentioned in the Tower of Babel. Instead, there is a poetic discussion among the godhead of how they must create different races and languages and scatter the people. The passage ends; and then the text returns to the flow of thought regarding the descendants of Noah.

Verse 9: *"From there the Lord scattered them over the face of the whole earth."* The tower being proposed was simply a large structure that would cause the people to *"make a name for ourselves and not be scattered over the face of the whole earth."* (Gen.11.4) They were in Mesopotamia, where large stones were scarce. They had to make bricks from clay.[5] Other ziggurat towers had been built. These were not huge edifices either.[6]

The practical take away from the passage is that God instructed humans to spread around the globe. He knew that sea levels were low, and groups could reach other continents by crossing land bridges. But as the planet warmed from the latest ice age, water levels were rise and soon cover the land bridges.

Whatever happened preceding and following a localized flood, humans eventually migrated. Many scientists point to DNA and the notion that indigenous North American peoples migrated from Asia. Many likely settled on the Bering Land Bridge for as many as 15,000 years until rising water levels forced them to continue to North America.[7] Thus, they preceded Europeans in the Americas by thousands of years. As the planet continued to warm from the last Ice Age, the land masses connecting the continents were again covered by sea water. Prior to being covered by sea water, the continents were connected by these land bridges. Once the land bridges were covered by water, the Earth was again *divided*. Continents were once again separated by water.

In the genealogy from Noah to Abraham, Genesis 10:25 mentions Peleg and that *"in his time the earth was divided;"* (also 1Chron.1.19) This puts Peleg on Earth a little more than 10,000 years ago when the land bridges were covered due to rising ocean levels as Earth warmed following the last ice age.

The genealogies of men from Adam to Abraham are likely partial lists, explaining why there are more than a few thousand years in the time span. A similar thing happens in genealogies of Jesus in the Gospels of Matthew (Matt 1: 1 – 16) and Luke (Luke 3: 23 – 38). Comparing the two lists of ancestors of Jesus exposes glaring differences. Believing these genealogies are comprehensive is a natural assumption on the part of a reader. In other words, the reader assumes that every ancestor between Abraham and Jesus is listed. However, a little math reveals these genealogies are nowhere near complete. Generations were less than 20 years. These genealogies cover a period of 2000 years. There would need to be over 100 names [generations] on these two lists to cover the entire 2000-year span. There are not. Since both lists skip a majority of ancestors, it is easy to understand why differences exist. Therefore, this is not a true contradiction bringing into question the validity of the gospels. Rather it is a recognition that lists of Hebrew genealogies do not necessarily include every generation.

When Bible scholars claim the Earth to be 6000 years old, they may fall into the same trap of assuming the genealogies from Adam to Noah and from Noah to Abraham in Genesis are complete with no skipped generations. Just like the genealogies in the gospels were not complete, neither are these unless there is another explanation. Looking to the notes in most study Bibles for the meaning of the Hebrew word 'abba' shown for Genesis 11: 10, one finds that the word can mean *father or ancestor*.[1] The mention that one ancestor *became the father* of someone can be interpreted as *became the ancestor* of someone. Or perhaps only the Jewish human line is discussed. After all, the first portion of the Sixth Chapter of Genesis is strange enough that latitudes for interpretation exist. For these and other reasons, the age of the Earth and the time of Noah cannot be determined by the Tanakh and Bible.

Prior to the flood in Genesis 6: 3, God declared his judgment would come on mankind in 120 years. Some think this verse states God would limit life spans of humans to 120 years; but this is not the correct interpretation. In context, it suggests the flood would occur in 120 years. After all, Genesis records Arphaxad, who was born two years after the flood, lived to be 403. Eber lived to be 464 years. After them, the ages continued to go down. Reu lived 239 years. (Gen 11: 20, 21)

The Tanakh and other sources mention persons living hundreds of years. This seems far-fetched. It is nice to stick to what is known and verifiable, such as the Fab Five and the clear plan involving the Messiah; but it would be appropriate to have some *what-ifs* regarding this idea of humans living to be hundreds of years old.

Here are some possible simple answers to the long-life claims. They range from "Aw, heck no. This portion of scripture is wrong and is not inspired."
To "This, too, is merely allegorical."
To "Maybe they should have put someone else in charge of producing the annual calendars."

But might claims of long life be scientifically possible? Perhaps, humans initially had higher concentrations of enzymes such as telomerase.[8] Telomerase is vital in the regeneration of body tissue. Avoiding meat and not being subjected to as many harmful rays as we are today, could prolong life. Earth has received a significant amount of harmful cosmic radiation from the Vella supernova in nearby space since some time in the past 40,000 years.[9] The timing of this supernova with an adjustment by God on certain enzymes could coincide with the dramatic change in average life length.

Long life could explain the question some ask about people living in Nod east of Eden. Genesis Chapter 4 describes that after Cain killed Abel, he moved to Nod, had a wife, and had children. Cain was a son of Adam. If Adam and Eve and their children lived for hundreds of years, they would have scores of children, who would each have many children as well. There would be thousands of people in the area. Some could have lived in Nod – hence, the ability for Cain to live there. Another explanation would be that humans living east of Nod were those from an orchestrated evolutionary process while Cain and Abel belonged to this newly created *God line*.

The point to be made is this: there are still many things we cannot know for sure. No one should have a problem scientifically with a localized flood in the Mesopotamian Region of the world at some time in the past 80,000 years. No one should be afraid of a possible evolutionary process which would require intervention by God at key points in the past four billion years. And no one needs to think that scripture and science deviate because someone had to fill unknown gaps with his or her bias. Archaeologists and other scientists will continue to discover.

Meanwhile, the plan provided by God to usher humans into an afterlife is well established in both scripture, history, and science. We can recognize who Jesus is and what this God of incredible life forms and the universe has done for us.

Chapter 10
Signs and Prophecies

The information contained in signs and prophecies from the Tanakh point conclusively to a Messiah who would eventually enter Jerusalem on the mission to die for mankind.

Over the years, many atheistic intellectuals have dismissed prophecy. Some make the argument that the Bible was written after the death of Jesus, and of course there would be prophecies and fulfillment written in such a book. The beauty of the way God established all this is what is gleaned when we listen to the Apostle Paul tell us that scripture is the Jewish Tanakh and not the entire Bible. The Tanakh was already in existence at the time of Christ. Thus, these prophecies and signs were all written centuries *before* Jesus arrived as a man. They are contained in the holy book of the well-established Jewish faith.

A priest with access to the temple at the time of the crucifixion could have read the scrolls describing the events taking place. It is why a spiritual ideology holding accountability to history and science concludes that Jesus is the Messiah and that his death and resurrection pave the way for us. It is important to mention that the birth of Christianity is the fulfillment of the Jewish religion.

Many signs and prophecies have been discussed already in this book. In 2000 BC Abraham broke bread and had wine with Melchizedek prior to the establishment of the covenant God made with humanity. Abraham gave a tithe to this person of the godhead – Jesus. The evening before Jesus was arrested and the events were put in motion for him to fulfill the sacrifice required in the covenant, he broke bread and had wine with his disciples.

When Abraham was walking his son up to the altar for the sacrifice, his son asked where the sacrifice lamb was. Abraham replied, *"God himself will provide the lamb."* On that occasion, God provided a ram whose horns were caught in a thicket. This was a sign of Jesus who would have his head in a crown of thorns at his crucifixion. And it was God himself, Jesus, who played the ultimate role as the sacrifice lamb.

In 1500 BC Moses was given instructions regarding the requirements of Passover. The blood from a perfect Passover lamb was to be applied to the tops and sides of the wooden doorways in the homes of Israelites captive in Egypt to avoid the death of their first-born sons. This act closely resembled the blood of Jesus on the wooden cross 1500 years later. God shared through Moses that in future Passover celebrations the lamb should be selected on the 10th day and sacrificed on the 14th day in preparation for the special Sabbath on the 15th day. This signified Jesus eventually coming into Jerusalem on the 10th day of Passover, being sacrificed on the 14th day of Passover, and being taken down off the cross before dark, because the 15th day was the special Sabbath.

Moses and – 500 years later – David both declared that the Messiah's legs would not be broken. When Jesus was crucified with two thieves, all three men needed to be taken off their crosses before sundown due to the special Sabbath. In order to accomplish this, the legs were broken to prevent a person from pushing up to get air. But Jesus was already dead, so his legs were not broken – fulfilling the prophecies.

The gospel of John records in Chapter 19 the scene at the crucifixion of Jesus. Matthew 27, Mark 15, and Luke 23 also record events. Jesus cried out, *"My God, my God. Why have you forsaken me?"* Soldiers had circled him and beat him to near death prior to nailing his wrists and ankles to the cross. They cast lots for his clothing.

Had a priest walked from where Jesus was being crucified into the temple to the scrolls comprising the Tanakh, the priest could have read all of what he just saw in the 22nd Psalm of David written 1000 years earlier - beginning with the words *"My God, My God. Why have you forsaken me?"* There, he would also read of a band of evil men encircling the Messiah and casting lots for his clothing and that his hands and feet would be pierced.

That same priest could then go to the scroll containing Isaiah's writings from 700 BC and read in Chapter 53 another detailed description of the crucifixion that was taking place. He would read about the Messiah being rejected by men, being crucified with criminals, and being buried in a rich man's tomb – which would occur in a few more hours. He could read of a Messiah who had to be pierced. Not only did Isaiah prophesy his being pierced, but so did Zechariah. This was done at the crucifixion to the lung to confirm that blood had mixed with water to fill the lungs and suffocate the person on the cross. Isaiah's prophecy declared that the Messiah would go through this willingly and not defend himself and that he would rise from the dead and see the light of life again. *"He would bear the sins of many and pour out his life to death."*

The priest could then go to the book of Deuteronomy [Chapter 21: 22 and 23] and read in the law that God would curse anyone who is hung on a pole – on a tree – such as Jesus. This is how God could justify cursing Jesus with all the sins of mankind. This priest could then make the connection to the Passover lamb and the blood on the wood in the pattern of a cross that would prevent the second death of men and women.

This same priest at the time of the crucifixion could then recall the procession that occurred four days earlier – on the 10th day of Passover. He could recall how Jesus entered Jerusalem on a donkey, the sign he was king according to the prophecy in Zechariah 9:9.

The priest could recall the cities of refuge established in The Law through Moses. [Numbers 35: 6 – 34] Persons accused of murder could run to a city of refuge, hold on to the horns of the altar of sacrifice, and plead for innocence. If it was determined the death was an accident, the individual had to remain in the city for the event that would wipe away any guilt and allow the individual to return home. The event to free him was the death of the High Priest.

This same priest might recall Psalm 118: 27 as chills went up and down his spine. David called the Messiah a high priest forever in the order of Melchizedek. Psalm 118: 27 was an invitation to those who would be alive 1000 years later when the Messiah would enter Jerusalem: *With [tree] boughs in hand, join in the festal procession up to the **horns of the altar**."* On the 10th day, the festal procession welcomed Jesus into the city.

It was now twilight on the 14th day of Passover. It was time for the Passover lamb to be sacrificed. The High Priest (the lamb) was dying on the cross, cursed with mankind's sins. Figuratively speaking, every soul of every human who lived or would ever live was grabbing on to the horns of the altar pleading for amnesty. Jesus spoke his final words: *"It is finished."* [John 19:30]. *"Father, into your hands I commit my spirit."* [Luke 23: 46] And then this High Priest died. Mankind was redeemed. The covenant made 2000 years earlier with Abraham was complete.

There are some prophecies that cannot be confirmed but are logically conceded. This includes the incredible unlikelihood that the Messiah would be born in the small town of Bethlehem. It is difficult to substantiate the claim in the Gospel of Luke [Luke 2: 1] that Caesar

Augustus declared a census and that residents had to make a pilgrimage to the town associated with their family line. This required Joseph to leave Nazareth and go to the little town of Bethlehem, where Jesus was reportedly born. The prophecy of the Messiah being born in Bethlehem comes from Micah 5: 2.

"But you, Bethlehem, Ephrathah, though you are small among the clans of Judah, out of you will come for me one who will be ruler over Israel, whose origins are from old, from ancient times."

But while historians have difficulty substantiating the census decreed, the logical conclusion is that it occurred. The physician, Luke, included it in his gospel. Had it not been true as people of the time would know, it would discredit the rest of what he had to say. Thus, logically it is a near certainty that Caesar Augustus declared such a census.

The birth of Jesus claimed in Bethlehem has additional significance to the prophecy. This was a town with many shepherds, who were among the lowest social class in Judaism due to their work with animals and having contact with sheep feces. It implies that after 13.7 billion years of preparation, God chose to be welcomed by the most lowly people on the planet. This confirms the notion that God loves the humble and resists the proud. Additionally, many of the lambs used annually in Jerusalem for the Passover were likely raised in Bethlehem.

Daniel prophesied the time of the Messiah written 500 years before Jesus arrived. In Chapter 9 of his book he intimates – using weeks as seven-year periods – how many years it would be from the decree to rebuild Jerusalem until the time of the Messiah. Historians using various estimates for the times of those decrees and using estimates of the precise year of Jesus death can confirm that the death of Jesus occurred within the window of legitimate times. While there is no way to confirm the exact fulfillment, we know it is at least extremely close and is why people around that time were looking for the Messiah.

One of the most descriptive and jaw-dropping prophecies involves the betrayal of Jesus by a close friend.   Matthew Chapter 27 records the actions of Judas. Judas agreed to betray Jesus for 30 pieces of silver. But after Jesus was captured, Judas became distraught and went to the temple to return the money to the high priest. The high priest would not accept it, so Judas threw the 30 pieces of silver back to the high priest in the temple. He eventually hung himself in a potter's field. The priests could not keep the silver, because it was blood money. Thus, they decided to pay the potter for his field, which many considered cursed since the betrayer of Jesus hung himself there.

These details of how the Lord was betrayed were prophesied in the book of Zechariah, Chapter 11: 12 - 13:

"I told them, 'If you think it best, give me my pay; but if not keep it.' So they paid me 30 pieces of silver. And the Lord said to me, "Throw it to the potter" - the handsome price at which they priced me! So I took the 30 pieces of silver and threw them into the house of the Lord to the Potter."

Of note, is that the JPS translation of the Tanakh to English uses the term *accountant* in place of *potter*. However, the Hebrew word is *yatsar* and is translated *potter* and not *accountant* in six other Tanakh passages, including Isaiah 41:25, *"he treads on rulers as if they were mortar, as if he were the potter treading the clay."* It is understandable that Jewish translators would question how a potter would be in the temple – hence, the translation to an accountant receiving the 30 pieces of silver.

This is one of the more impressive prophecies. How were the coins – the ransom paid for betrayal of the Lord – going to be thrown into the temple to pay some potter?

Another sign involved Moses after he led the captive Israelites out of Egypt. This large group of people were in the desert, needing water. Many were angry at Moses and his brother Aaron. Exodus 17: 4 records

Moses crying out to the Lord, *"What am I supposed to do with these people? They are almost ready to stone me?"* Exodus 17 continues, *"The Lord answered Moses, 'Walk on ahead of the people. Take with you some of the elders of Israel and take in your hand the staff with which you struck the Nile, and go. I will stand there before you by the rock at Horeb. Strike the rock, and water will come out of it for the people to drink.' So Moses did this in the site of the elders of Israel."* [Ex 17: 5 – 6]

A year later the group was back at the same location. The people again grew angry and were without water. Again, Moses and Aaron were frustrated and turned to the Lord. Numbers 20: 6 – 13: *"Moses and Aaron went from the assembly to the entrance to the Tent of Meeting and fell facedown, and the glory of the Lord appeared to them. The Lord said to Moses, 'Take the staff, and you and your brother Aaron gather the assembly together. Speak to that rock before their eyes and it will pour out with water. You will bring water out of the rock for the community so they and their livestock can drink.'*

So Moses took the staff from the Lord's presence, just as he commanded them. He and Aaron gathered the assembly together in front of the rock and Moses said to them, 'Listen, you rebels, must we bring you water out of this rock?' Then Moses raised his arm and struck the rock twice with his staff. Water gushed out, and the community and their livestock drank. But the Lord said to Moses and Aaron, 'Because you did not trust in me enough to honor me as holy in the site of the Israelites, you will not bring this community into the land I give them.'"

This seems harsh on God's part. Moses risked and accomplished much. Why did him losing his temper prohibit him from entering the promised land with the congregation? One issue may have been Moses stating, *"Must we –* implying Moses and Aaron *– bring you water out of this rock?"* This may suggest Moses took credit for the water. But there is a far more important issue.

Like Abraham meeting Melchizedek and breaking bread and sharing wine before the establishment of the covenant, the interaction with the rock was to be another sign of the Messiah, Jesus. The Messiah would be struck down once for mankind. Hence, initially God instructed Moses to strike the rock once. With the second interaction at the rock, God commanded Moses to speak to the rock and water would come. This is a foreshadow of how the sacrifice and resurrection of Jesus would allow men and women to pray to God, and God would grant his Holy Spirit to live within them. But Moses failed to follow through with this order from God, destroying the poignant symbolism.

God had to make a big deal out of this to preserve this important sign. Paul stated 1500 years later that the rock accompanying the Israelites in the dessert was Christ Jesus. [1 Cor 10:4] Notice in the first instance of Moses and the rock recorded in Exodus that the Lord stated, *"I will stand before you by the rock at Horeb."*

The plan of God was precise, with multiple signs, actions, and prophecies pointing to it. From a mathematical perspective, there is no way to expect these fulfilled prophesies are mere random coincidences – not when they are found in the one most commonly accepted scripture on Earth and when they contain so many vivid details of the Messiah, his purpose, and what he would endure. It suggests the simplest and most logical explanation for the presence of human beings on Earth.

Knowing the existence of the Tanakh and its incredible detail of events that would happen centuries after it was written, we know that humans have been contacted by a being who exists beyond our space-time dimension. With this condition established and applying conditional probability to the explanation for the existence of humans on Earth today, the strongest scientific conclusion is that this triune God created the universe and we humans. The details of how He used evolution and natural progression along with moments in time where he intervened with mass extinctions and mass repopulations cannot be precisely known.

Knowing how rare it is to find a planet with Earth's physical characteristics and beneficial properties and knowing how close Earth came at various times to becoming sterilized, it is not rational to conclude that human life on earth is a product of random chance.

Another important point should be made. Usually more than once per year, NASA locates a new planet in a habitable zone in nearby space that *might be habitable.* This does not mean it has many of Earth's characteristics. It merely implies it is located in the habitable zone of its star(s). It may be small and rocky with water, but it does not imply there is life, and it does not imply that animals could be transported and survive there.

Recall that Naturalism is what is taught as science in American public schools. But Naturalism is a philosophy and not pure science. Pure science makes no assumptions. However, Naturalism assumes that no influence can exist outside our space-time dimension. The argument against naturalism is strong and damning.

There is currently no explanation for the origination of self-replicating material on earth. Naturalism cannot offer explanations for mass, sudden repopulations that followed significant extinction events and seemed to anticipate future environmental conditions. Some ideas still promoted in classrooms have been discredited for years. Such is our existence in a time when so much information is available. We as humans – regardless of intelligence levels – feel safe and affirmed when we belong to a large group of like-thinking people. It is yet another example of the human-nature dynamic created by God. Can an individual get past the stumbling blocks of human nature?

Coming to grips with the reality of this life requires humility and integrity. And it requires one more component – a spiritual component. People can read a book like this, read of the fulfillment of prophecies, see the frailty of Naturalism, and be shown the plan and the steps to acquire this gift of life-after-death. Yet, many will not process it all. We are complex beings with emotional, intellectual, and spiritual components. The humility and integrity must occur on all three levels to have the eyes of our understanding open to such a brilliant plan and such a brilliant and loving God.

Much like financial and business situations can be viewed from the standpoint of a game, so can life itself with its true purpose. Solving life's meaning – and thus winning the game – requires effort and virtue. And often the revelations and understandings require pain and suffering.

We are good and we are bad. We are wise and we are ignorant. We are powerful and we are susceptible. Just like the physical universe and physical make-up of life are beautifully complex and precise, so is our very existence from birth until death. And so is the plan to redeem the lives of men and women and provide the path to the afterlife. The interactions and experiences in our lives have great importance.

There is no place for bigotry as God is the God of all people groups. There is always a place for humility, integrity, love, valor, and courage. We should concern ourselves with the welfare of others by developing empathy and carefully considering their plight. The world needs less judgement and more followers and emulators of Christ, the greatest example of a life lived with purpose, care, and concern for others.

Chapter 11

The Spiritual Ideology

These pages contain the foundation of a spiritual ideology accountable to Scripture, Science, and History. The concepts are in primarily chronological order. Transforming our lives for the better is a process that takes time. Key in the process is the attempt to be life-long learners – continuing to question, listen, reason, and insist on accountability.

Quotes here tend to be from the New International Version of the Bible. No translation is a perfect translation from the original Hebrew of the Tanakh or the Greek of the New Testament. In fact, at various times, well-meaning translators injected their bias into their translation. It is important to attempt to hold accountability to the original Hebrew and Greek and the likely interpretation.

Additionally, The Tanakh [Christian Old Testament] is spiritually discerned and can be veiled – requiring diligence and humility. (2 Corinthians 4. 1 – 8). We pray for wisdom and understanding. It is the Holy Spirit who opens our eyes spiritually. But be careful. This is not a license to believe anything we assume is from God. If our ideas and interpretations cannot mesh with scripture and the teachings of Paul and Jesus, then why would we assume that they are insights from God?
2Cor: 4: 1 - 8; John 7: 37 – 39; 1 Corinthians 1: 18 – 2: 16

What follows are several key doctrinal ideas. This is not an exhaustive list – meaning there is more involved. It follows a primarily chronological order.

Triune God: In the Tanakh, the Jewish holy book for a Jewish religion that believes in one God, it is impressive to see the quote in Genesis [1:26]] *"Let us make man in our image."* This implies one God with more than one personality. In John [John 17: 1 – 5], Jesus is praying to the Father God and discussing how they made plans before the beginning of the world. The Holy Spirit is the third member of the Trinity. In Acts 5: 3,4 lying to the Holy Spirit was lying to God.

> *Gen: 1:26; John 17: 1 - 5; John 1: 1,2; John 11:41,42; Matt 28:19;*
> *2 Cor 13:14; Acts 5:3,4*

Space-time dimensions: In 2Tim 1:9,10, Paul tells us that, *"The grace given to us in Christ Jesus was given to us before time began."* Genesis 1:1: *"In the beginning, God created the heavens and the earth."* Many scientists as recently as the beginning of the 20th century still assumed that the universe was infinite and static. Many religious ideologies believed this as well. But the Tanakh accurately acknowledged that our space-time universe began at some point in the past. And as the author in Hebrews [11: 1 - 3] accurately reported, the universe was made from that which is unseen. Subatomic particles cannot be seen with the naked eye or with powerful microscopes, but God created with such material not visible.

God must exist in at least two more dimensions of space and two more dimensions of time compared to the space-time dimension of our universe to be in control of both our physical realm and a spiritual realm. The prayer Jesus made to God when Jesus was on earth is also an indication of this. John 17: 4, 5: *"I have brought you glory on earth by completing the work you gave me to do. And now, Father, glorify me in your presence with the glory I had with you before the world began."*

> *Gen: 1:1; Hebrews 11:1 - 3; John 17: 4-5*

The Genesis account of creation, space, and Earth describes six long periods of time – one of which is more than one billion years long. The Hebrew word for day is yom [or yowm] and can mean period of time.

108

The passage interprets itself with various contextual clues, including how there was *"evening and morning"* of each day. But on the seventh day [Chapter 2], there is only evening and no morning recorded. Hence, we are still in the 7th day of creation – 7th long period of time. God is still very active but has rested from creating new things. Instead, species adapt and change. They evolve over time.

There is little reason to doubt Earth being 4.54 billion years old and the known universe being 13.7 billion years old.

Gen: 1:1 – 2:3; Psalm 90: 1 – 4

Unlike other animals, humans are part spirit. This is what was meant when God said, *"Let us make man in our image."* We have a spiritual component in addition to a physical body. When we are *"born-again"* spiritually, it implies we are born into a heavenly realm that exists on Earth now and will continue for eternity in the presence of God.

Genesis 1: 26; Luke 17: 22 – 37

God exists in a realm that includes and is outside of our space-time dimension. This is why God described himself to Moses as *"I AM"*. And this is why the author of Hebrews declares Jesus Christ as *"the same yesterday, today and forever."* Nothing gets past God. He cannot be fooled by beings he created as a result of this. This implies more of the doctrines that follow, and this is how **God is in control while humans have a free will.**

The death that came to mankind as a result of Adam's sin was a spiritual death. Many animals died prior to human presence. It provides the fossil fuels for today's civilizations. Humans were not created perfect. They were created with good and bad tendencies that make it possible – for those who fight human tendency and seek with humility and integrity – to determine that we are flawed and yet God loves us so much that he prepares an afterlife and paid for the passage. Death physically was inevitable in this finite universe that would eventually burn out and die if

left to itself. Human flesh and the universe were not created for eternity. The human spirit was.
Exodus 3: 14; John 8: 58; Hebrews 13.8

God created a spiritual realm of beings before He created humans. Jesus shared what he saw long before he came to Earth as a man. *"I saw Satan fall like lightning from heaven."* [Luke 10:18] We are told in John 16: 11 and John 12: 31 that Satan is the prince of this world and that we wrestle with principalities and powers in the heavenly realm [Ephesians 6: 10 & 11]. In scripture, there is an account of an angel being sent to answer a prayer from Daniel. The angel was intercepted by the prince of Persia – a spiritual being: an evil spiritual being. Finally, the archangel Michael came and freed the angel to get to Daniel.   **Conclusion: God created beings he knew would turn and do evil.** This was part of his purpose for this life and the need for us to exist in a universe with pain, suffering and evil.

Hell is a real destination prepared for the Devil and his angels, and humans old enough to make a decision may ultimately decide to join Satan in this place of torment. God is not willing that any humans end up there, but many will. It is part of the brutal necessity of this short life in preparing an eternal life of optimal interaction. People in the afterlife will have grateful hearts and be given a new body not prone to simple negative human traits. The combination of beautifully committed and transformed souls and an existence in a more dynamic space-time existence makes heaven more incredible than we can fathom. The precise details of hell are not fully known. Jesus tells us to fear it and to respect [fear] God who will send us there if we ignore His provision.
2 Luke 12: 4, 5; Matthew 25: 31 – 46; John 3: 1 – 21

Recordings of past individuals and their lives and struggles help us today to recognize the good and bad in us. The majority of great men and women recorded in the Tanakh made major mistakes. Yet, God used them and created signs from their lives to help seekers realize the validity and inspiration of his plan to come to Earth and be sacrificed (killed) for our sins. Abraham's impatience produced two blessed people groups that have warred against each other throughout the centuries. Moses killed a man. David committed adultery and then devised a plan that would cause the incredibly devoted husband of this woman to be killed in battle. Saul [later Paul] hunted early Christians – imprisoning some and killing others. None of us want to do bad. We should be careful not to assume we are right. When God's followers get reckless, it can lead to enormous consequences. Nothing should be taken for granted. It is healthy to admit humanness and to focus on doing good whenever possible. The wonderful recognition of God's grace, love for us, and desire to use us to better the lives of others is to recognize that we cannot lose our salvation when our heart is in the right place in making a commitment to him and attempting to persist.

Ephesians Chapter 2

The recording in the Old Testament of a massive flood gets many different reactions from people. Most assume it is too far-fetched and must be a symbolic tale or evidence that scripture is ridiculous. Others believe it is a literal depiction, that it covered the entire planet to the tops of the highest mountains; and some believe it happened less than 6000 years ago. But as eluded to earlier in this book, the Genesis account of a flood interprets itself. It was not global. It was regional. This is in line with the account in Genesis 7 & 8. It is also in synch with Psalm 104, which suggests that – after the initial emergence of land on Earth billions of years earlier – never again would the seas cover the tops of all mountains.
See Chapter 9 in this book.

Psalm 104: 1 – 8; Genesis 7 & 8

Jesus, also known as the heavenly high priest, Melchizedek, came to Earth and met Abram [later called Abraham]. They broke bread and shared wine, and Abram gave a tithe to Melchizedek. This happened prior to Abraham being willing to sacrifice his son: before the establishment of God's covenant with man. Centuries later, Jesus would again break bread and share wine – this time with his disciples – prior to the fulfillment of the eternal covenant with God through the sacrifice of Jesus.

Genesis Chapter 14; Matthew 26: 26 - 29

God established a covenant with Abraham. God found a man so devoted that the man was willing to sacrifice his own son to God. This faith was credited to Abraham as righteousness, and the covenant was sealed between God and man. The promises and blessings made to Abraham can be ours through faith in God's plan alone – faith in Jesus and what he accomplished. In the recording of the events surrounding Abraham taking his son to the alter to be sacrificed, there are many clues pointing to the Messiah [God, the Son] and what he would suffer and accomplish 2000 years later. Abraham's son asked Abraham where the lamb was for the burnt offering. Abraham replied, *"God, himself, will provide the lamb."* This statement had double meaning. First, it implies that God himself will be the sacrifice lamb in the form of Jesus 2000 years later. Second, on that day a ram was found with its horns caught in a thicket. It was sacrificed. It was a foreshadow of Jesus having his head in a crown of thorns at the time of his crucifixion.

Genesis Chapter 22

Jesus completed the covenant by being beaten and hung on a cross.

Galatians 3: 13, 14; Romans 10; John 19

God used Moses to deliver his people from slavery in Egypt. The final plague that got the Hebrews released was the first-born son dying during the night. The only way for death to pass over the home was to sacrifice

a perfect Passover lamb and put some of its blood on the sides and top of the wooden door frame – a foreshadow of Jesus, the Passover lamb, having his blood placed on the wooden cross to allow those who put faith in God's provision to have the second (eternal) death *pass over* them.
Exodus Chapters 11 & 12

God gave the Law through Moses. [see chapter 7 in this book] The Law accomplished the following:
It set a standard no human could attain.
Humble individuals could recognize this and investigate how to be considered acceptable in God's eyes.
Self-centered, egotistical people would fall into the trap of convincing themselves they could live by the Law and be acceptable in God's eyes.
Others would think that the law is so ridiculous (which may seem true) that the notion of a god being behind it would be nonsense.
It established laws allowing God to sacrifice God the Son and curse him with the evil done by anyone during a lifetime. *"Cursed is anyone hung on a tree."*
It provided prophecies and signs that would confirm God's plan, existence and faithfulness.
> *Deut 21: 22 - 23*
> *The Law is the set of rules and regulations dispersed throughout the first five books of the Tanakh [Old Testament]*

God influenced men from Job prior to 2000 B.C. through some of the prophets in the centuries preceding the arrival of Jesus to write the scriptures, the Jewish Tanakh.
Signs, prophecies and God's control of history would provide the proof of His existence and His plan and give men principles from which to live and from which to anticipate the Messiah's arrival and provision for

eternity.
> *2 Peter: 1: 19 – 21*

The Jewish Tanakh is the "inspired scriptures" to which the Apostle Paul referred.
> *2 Timothy 3: 14 – 17; 2 Peter: 1: 19 – 21*

Jesus arrived and fulfilled many signs and prophecies
He is God and human according to many indications. John describes this well in John Chapter 1
His mother, Mary was a virgin. Jesus was conceived of the Holy Spirit. This was prophesied in Isaiah 7:4 and recorded in the Gospels of Matthew and Luke.
His death and resurrection provide the portal for all to the afterlife
His act allowed the Holy Spirit to dwell in the hearts of believers
and not just in the Holy of Holies of the temple or tabernacle.
He taught his disciples for three years prior to his death. After
his death, the disciples traveled around the known world,
spreading the news of what was done for Jews and for
Gentiles (as was prophesied).
> *John Chapter 1; Isaiah 46.9; Matt 28.19*

Before the beginning of time, God devised a plan allowing those who accept the grace of Christ to dwell with Him in a superior realm referred to as heaven. He influenced Moses to write the law that *"Anyone hung on a tree [pole] is cursed."*
Paul explains how Jesus was cursed on the cross in this manner and is our provider in the afterlife.
> *2Tim 1:9; Deut 21: 22 – 23; Gal 3: 10 - 14*

Jesus rose from the dead as prophesized by Isaiah and others.
> *Isaiah 53: 9 – 12; Psalm 16; I Cor 15: 12 – 20;*

Christianity is the fulfillment and continuation of Judaism. The Tanakh states that Gentiles along with Jews would reap the reward of the covenant.
Psalm 22: 27 – 31; Isaiah 49: 6

Writings in the New Testament record the fulfillment of scripture and teachings of Christ and the Christian faith as the completion of the promise made to man by God.

God equips us with everything we need if we allow Him to do so. Even faith is a gift from Him.
Ephesians 2: 8; Ephesians 6: 10 – 18

As promised in Scripture, the Holy Spirit now abides within believers as opposed to only dwelling in the tabernacle of God in the days of Moses.
He nudges us in the right direction. He gives us wisdom when we need it. He comforts us. He was promised to us by many prophets, and the delivery was made possible by the sacrifice of Jesus.

The Holy Spirit [The Comforter] was promised to the disciples of Jesus and to all future followers. The early disciples were told to pray and wait in Jerusalem until he came. The Book of Acts records His coming on Pentecost. It stated that He came like a rushing wind. [Acts Chapter 2] And when the disciples took to the streets to share what happened, they spoke in different tongues – different languages – so that foreigners in Jerusalem heard the message of Jesus and God's plan in their own native tongue.

There are powerful gifts the Spirit gives to different people to make the Church function more efficiently and effectively in its mission to share the gospel message with as many people as possible. [I Cor 12]

A word of caution: some churches can take the actions of the Holy Spirit to extremes – and some seem to pretty much ignore Him. There are some well-intentioned people and some manipulative people who can create a complicated, emotional set of practices involving the Holy Spirit. Be careful when people say, "God told me this," or the "Spirit is leading me in this direction." Other than searching scripture and the New Testament to see if it supports the claims of such individuals, there is little available in the way of accountability to test for trustworthiness. We should practice humility and state we *believe* God to be leading us.

While the Holy Spirit himself is the most powerful force in our everyday lives, teachings surrounding the Holy Spirit can be some of the most destructive in the world today. We should always err on the side of love and accountability to scripture and biblical teaching. Just because people we know may be good and kind in general, does not imply that they have a powerful take on what the Holy Spirit is suggesting for their life or ours. We should respect the spiritual realm and recognize how fallible we are in attempting to interpret signs in our lives and the correct actions to be taken.

The good news is that – in addition to us accepting the gift of eternal life – God provides the Holy Spirit to guide us and take care of us in our daily lives.

> *John 7: 37 – 39; John 14: 15 – 17;*
> *Ezekiel 36: 25 – 27; John 16: 13 – 15; Romans 8: 9 - 16*

Paul – after his remarkable conversion from being the persecutor, Saul of Tarsus – became the intellectual who explained the scriptures, the plan, and key details and perspectives to instruct the billions who would eventually seek Christ.

> *Acts Ch 9; all of Paul's letters [Romans, 1&2 Corinthians,*
> *Galatians, etc.]*

Prayer is a powerful way for believers to connect directly to God through the Son. And Jesus prays continually for his believers on Earth. He taught us a model for prayer that includes the components of how we should communicate with God and recognize our relationship with him

and dependence on him. See the Lord's Prayer [Matthew 6: 9 – 13]
> *Romans 8: 31 – 35; Hebrews 7: 25; Matthew 6: 5 – 15;*
> *1 Thessalonians 5: 16 - 18*

As God, Jesus knows our human condition well. His teachings, often shared in stories or parables, help us to understand important perspectives on life.
He taught the story of the Good Samaritan (which seems to elude many today). In it, he portrayed how many self-righteous people avoid looking out for others. Yet, a Samaritan – from a group of people looked down upon in the culture of that time and area – was the one who showed a tremendous capacity to care for a total stranger.
> *Luke: 10: 25 - 37*

This life is designed to have tribulation and struggle; and we grow as we persevere through it wisely.
God made no mistakes because he is in the future and the past simultaneously and so cannot be fooled. He made a complex universe with pain, suffering and evil playing an important dynamic in getting the attention of people and in displaying the essence of love as people rise above challenges to be compassionate and dedicated to others.
> *John 16:33 [in fact, John 16: 17 – 33. He is speaking*
> *to his immediate disciples] James 1: 1 – 18*
> *[In fact, all of James Chapter 1 is beneficial]*

Baptism is an important display of one surrendering his or her earthly, carnal being and then accepting the gift of rebirth into a life led by the Spirit. We are to be baptized in the name of the Father, Son, and Holy Spirit.
> *Matthew 3: 13 – 17; Acts 2: 37 – 41; Matt 28: 19*

The love of God, exemplified by Christ, and the fruits of the Spirit are now the spiritual law of the day and accomplish what the law given to

Moses could not. A person's ability to practice humility and integrity despite our human tendencies plays the greatest role in us searching for the reality of this life, the existence of God, the recognition of our condition, and the plan and the gift paid for by the grace of God. It adds great purpose to this life and is therefore the source of great joy and fulfillment.

Galatians Chapter 3; Galatians 5: 16 - 26

Jesus summarized the law and the prophets – how we are to live – in the simple statement, *"Love the Lord your God with all your heart and with all your soul and with all your mind. This is the first and greatest commandment. And the second* [greatest] *is like it. Love your neighbor as yourself."*

Mark 12: 28 - 31

Each person's heart is the determining factor in salvation and in effective service through love rather than obligation.

1 Samuel 16: 7; Luke 17: 20, 21; Matthew 6: 19 –21;
Matthew 25: 31 - 46

Confessing Jesus and loving God through one's actions are a display and litmus test for salvation.

Romans 10: 8 – 13

Paul suggests in Romans 2: 12 – 16 that some will enter the kingdom of heaven without hearing the gospel. They will be judged by how they followed their heart. It is important to note that it is still the sprinkling of the blood of Jesus that saves them.

Faith and love (and the good works that come from these motivations) are the desired works as opposed to when people do good things out of obligation or selfish gain. God loves it when – as much as possible – our good works are done for others with few, if any, strings attached.

Ephesians Chapter 2
***Paul summarizes much of what we have discussed in this
Chapter of Ephesians.
1 Corinthians 13 describes many aspects of love.

There is a very organized evil spiritual realm. While accounts in the New Testament refer to instances of demon-possession, there is a far greater power at work on the planet. In Paul's letter to the Ephesians, he states we wrestle against principalities and powers. God created these beings to make the trial of this life complete. Subjecting humanity to this beautiful, yet brutal life places humans in sometimes desperate situations. When we get to the next life, we will recognize the benefit of experiences here. Emerging through what can be described as *trial by fire* will produce dynamics in an afterlife of free-will beings not attainable otherwise.

The combination of these spiritual powers working in governments, schools, churches, military, the financial sector, and other groups along with the flaws and inclinations of human nature make the dynamics of life tenuous. Countries can move from relatively healthy cultures to unstable ones in short periods of time. Promotions of thought can be organized by these spiritual forces. Ignoring their existence has consequences.

Ephesians 6; Daniel 10

There will be a day of judgment spiritually in which God will separate the faithful from the non-believers. There will be horrifying agony for those who will be cast to hell with the fallen angels.

Matthew 25: 31 – 46

The afterlife will be a place God is preparing for us. We only arrive there by accepting God's redemption through the sacrifice of Jesus, the Messiah. It is a new heaven and new Earth and a new body in a greater space-time dimension. Every aspect of it will be remarkable – from relationships, to physical dynamics, to fulfillment and being in the presence of God. *"'What no eye has seen, what no ear has heard, and what no mind has conceived' – the things God has prepared for those who love him."* [1 Cor 2: 9]

John 14: 1 – 3; Rev 21: 1

The Earth and this universe will ultimately be destroyed. A new space-time dimension with a new Earth will emerge. When a person dies, he or she is ushered to this different space-time dimension which already exists. It is impossible for any of us to conceive it due to its multiple dimensions of space and time. We as spiritual beings created in God's image will reap the reward of such an incredible life through the faithfulness of Jesus and our acknowledgment of him.

Luke 21: 33; Matthew 5:18; Prophecy in Revelation 21:1

Many signs and prophecies are highlighted in Chapter 10.

PLENTY OF ROOM FOR DISAGREEMENT:

There is much room for disagreement among religious bodies regarding many of the details of life and how to live. Therefore, there are some doctrines not discussed. The encouragement is to stay true to literary, historical and scientific integrity. We should stay true to the spirit of the law of God and the love He promotes. Our greatest responsibility is to help others recognize who God is and what He can do for everyone. Our inaccuracies can push people away from God. We should study well and struggle well.

Appendix

How to Become a Christ Follower

How to become a follower of Christ and accept the gift of eternal life and of the Holy Spirit:
God's plan to redeem us all and allow us to join him in the afterlife is extremely precise. It requires laws and signs and prophecies and precise actions. However, when it comes to His interaction with us, grace and mercy are the best descriptors. God looks on our hearts and promises that, though this world and life have their tribulations, all of it will work toward good in our lives and others as we dedicate ourselves to Him and accept the gifts we could never earn on our own.

Jesus and others like David help us to recognize we do not need to get caught up in formality. Many with respect of God practice elaborate ritual. Others are pretty lose. Sincere and loving interaction with God is as varied as our individual personalities. We should have great respect and a bit of fear in making sure our conversion and devotion are genuine. There can be much peer pressure, guilt, and confusion depending on circumstances.

Paul's letter to the Romans is perhaps the best single source to read repeatedly as one grows. It is a healthy balance of devotion and grace and understanding. Our mind and heart need to be recalibrated and synchronized so we can better hold our emotions and the negative influence from other sources in check. Rely on Romans more than any other source, including a pastor. Remember, there is a battle for souls and a realm of negative spiritual influence. Frankly, most Christians fail to realize how influential and subtle it can be.

In my opinion, many churches and Christians assume that, because we are on God's team and have the Tanakh and Bible, we are following the Holy Spirit well and must be accurate with all our views. Ideas from iconic leaders are often quick to be embraced. This can be dangerous. I am extremely concerned how many groups of Christians are accepting and promoting serious misinformation threatening our health, our political process, and our civility.

Therefore, my suggestion is for us to trust Paul to guide our growth and help our mind and heart develop in a healthy way. Romans is the go-to source for appropriate mindset.

If we are still wrestling with Jesus as Messiah and His act paving the way for us to enter the kingdom of God, here is a very good prayer to offer up to God:
"Hey God: please help."

Seriously. Learn from the start that prayer does not need to be some lavish, poetic phrase as if we lived in a place and time where Old English is necessary. Most have an opinion, but I cringe a bit when I hear prayers with all the latest buzz words and common phrasings like the person was entered in some prayer competition.

If you want a prayer to pray when you are ready to commit your life to God, feel free to use one I created for such an occasion. Hebrews 4: 14 – 16 reminds us how awesome Jesus is and how well he relates to us.
"Therefore, since we have a great high priest who has gone through the heavens, let us hold firmly to the faith we profess. For we do not have a high priest who is unable to sympathize with our weaknesses, but we have one who has been tempted in every way just as we are – yet was without sin. Let us then approach the throne of grace with confidence, so that we may receive mercy and find grace to help us in our time of need."

Here is a possible prayer:

Dear God:

I recognize my human condition and that I act well, and that I act poorly. I believe in your plan that can allow me to spend eternity with you and to be led and cared for by you in this current life. So it is by faith that I confess I fall well short of perfection. My hope is not in my ability to live appropriately at all times. My hope is in your fulfillment of the Law and of your covenant with us through Abraham and Jesus. I ask your forgiveness for my infractions, and I accept the provision made through the sacrifice of Jesus. I dedicate my life to you.

Please give me your Holy spirit to guide, nudge and comfort me through the rest of my days here. Please help me grow in wisdom and help me make a difference in the lives of others. Protect me and help me be strong in a world that is both beautiful and dangerous. Help me not to be overly burdened with guilt or other emotions that can stunt my growth and joy in this life. Help me to find a healthy balance between work, service and leisure. Help me to find healthy sources with which to grow. Please help me to give my time and money to important causes and to take responsibility for people around the globe who are suffering. I am willing to work hard. Please provide the opportunities and means to meet my financial needs and obligations.

I know that life can get hard. Help me to struggle well in those times. You obviously know me better than I know myself. Please direct my path so I can use the gifts you have given me and so I can receive the godly desires of my heart. Thank you for what you have done and are doing for me. I ask all of this in the name of the Son and Messiah, Jesus Christ.

If you are worried about messing up now or in the future, don't worry – you will. You are human. Read Romans. Read the four Gospels and the rest of the New Testament. {Good luck with The Book of Revelation... ??? Just glean general ideas. Most people have silly notions of its specifics.} Go to Old Testament passages referenced while reading the New Testament. Re-read the Spiritual Ideology and verses I list. After this, read the Psalms and Proverbs. Then read the first five books of The Old Testament {Tanakh}. Finally, read the remainder of the Tanakh {Old Testament}. A respected Study Bible can help.

Balance work, service, and leisure. Have determination and faithfulness. Have fun. Life is exciting and challenging enough. Avoid needing it to be spectacular and overly emotional. Doubt every conspiracy theory you hear. If not trained how to vet them, stick to what you can know and do not get mired in them.

Paul compares the Christian life to a race – a very long race. Some days it is easy. Some days it is hard. Sometimes we reflect and realize we are experiencing great faithfulness and effectiveness. Other times we realize we have not been very productive.

Try not to let fear or guilt take hold of you. Seek first God's kingdom, and so much will be provided for you. But also know that some of us will suffer greatly. Do not get bitter. Always put this short life into the perspective of the life that is to come. And it will come. God is real as are His promises. While it is appropriate to be concerned with our personal circumstances, we should routinely consider the plight of others and how we might be of assistance and encouragement. Giving with little concern for our gain or reward is often quite fulfilling. Collectively, our Great Commission is to help others recognize the plan God provides and fulfills.

Every person matters and should be respected. If we cannot truly learn something valuable from another person, we may not be looking hard enough or our ignorance and EBS may be getting in the way.

We should enjoy the good times and treat ourselves and those we love well. Yes, we should work and serve. But we should find time to have fun, and to experience this planet, time with loved ones, and the things that bring us joy. Struggle well through all of life's challenges.

The afterlife will be absolutely incredible. See you there

Move to Reality

References

Chapter 1:

1   Daniel Kahneman. Thinking, Fast and Slow. Farrar, Straus, and Giroux, 2013. Pages 39 – 49.

2   Sophia Smith Galer. "The Accidental Invention of the Illuminati Conspiracy" BBC.com BBC Future August 9, 2017 https://www.bbc.com/future/article/20170809-the-accidental-invention-of-the-illuminati-conspiracy

Chapter 2:

1   Dr. Fazale Rana. "Do Self-Replicating Protocells Undermine the Evolutionary Process?" Reasons to Believe. November 12, 2015   http://www.reasons.org/articles/Do-Self-Replicating-Protocells-Undermine-the-Evolutionary-Theory>

2   Ruth Marlaire. "NASA Ames Reproduces the Building Blocks of Life in Laboratory" NASA. Solar System and Beyond, March 3, 2015
        www.nasa.gov/content/nasa-ames-reproduces-the-building-blocks-of-life-in-laboratory
3   Pearce, Ben K.D. and Pudritz, Ralph E. "Meteorites and the RNA World: A Thermodynamic Model of Nucleobase Synthesis within Planetesimals." Astrobiology 16 (Nov 16): 853 - 872

4   Ibid

5   Peter D. Ward and Donald Brownlee. Rare Earth: Why Complex Life

Is Uncommon in the Universe.
Copernicus Books, 2004. Page 98.

6   Viviane Richter. "The Big Five Mass Extinctions" Cosmos
Paleontology
cosmosmagazine.com/paleontology/big-five-extinctions

7   Hugh Ross. Improbable Planet: How Earth Became Humanity's
Home. Baker Books, 2016. Page 169.

8   Ibid, 167, 168

9   Hugh Ross. Why the Universe Is the Way It Is. Baker Books, 2008.
Pages 44 – 47.

10   Peter D. Ward and Donald Brownlee. Rare Earth: Why Complex Life
Is Uncommon in the Universe.
Copernicus Books, 2004. Page 42.

11   Hugh Ross. Improbable Planet: How Earth Became Humanity's
Home. Baker Books, 2016. Page
38,39

12   Ibid

13   Ibid, 112

14   Kurt O. Konhauser et al. "The Archean Nickel Famine Revisited,"
Astrobiology 15 (October 15): 804

15   Hugh Ross. Improbable Planet: How Earth Became Humanity's
Home. Baker Books, 2016. Page 123.

References

16   The Drake Equation. "What do we need to know about to discover life in space?" SETI Institute
   http://www.seti.org/drakeequation

17   Mike Wall, Space.com Senior Writer, "NASA Finds 1284 Alien Planets, Biggest Haul Yet, with Kepler
Space Telescope" Space.com Science & Astronomy, May 10, 2016
   http://www.space.com/32850-nasa-kepler-telescope-finds-1284-alien-planets.html

18   Ibid

19   Nola Taylor Redd. "New Moon-Formation Theory Also Raises Questions About Early Earth"
Smithsonian.com. Science-Nature, August 12, 2016
http://www.smithsonianmag.com/science-nature/new-moon-formation-theory-also-raises-questions-about-early-earth-180960077/

20   Hugh Ross. Why the Universe Is the Way It Is. Baker Books, 2008. Pages 80, 81.

21   Ibid

22   Jesse Emspak. "The Moon Was Formed in a Smashup Between Earth and a Near Twin: But solving
one puzzle of lunar origins has raised another linked to the abundances of tungsten in the
primordial bodies" Smithsonian.com, April 8, 2015
http://www.smithsonianmag.com/science-nature/moon-was-formed-smashup-between-earth-and-near-twin-180954915/

23   Elizabeth Howell. "Reference: Shoemaker-Levy 9: Comets Impact Left Its Mark on Jupiter,"

Space.com Science & Astronomy, February 19, 2013.
http://www.space.com/19855-shoemaker-levy-9.html

Chapter 3:

1 Strong's Exhaustive Concordance of the Bible, Updated and Expanded Edition. Word 3117 [yowm].
    Peobody, Massachusetts: Hendrickson Publishers, Inc, 2007

Chapter 4:
1   Herodotus, Hist. i.128; iii. 132.2, 159.1

Chapter 5:
1 Matt Williams. "A Universe of 10 Dimensions" Universe Today, November 7, 2016
    https://www.universetoday.com/48619/a-universe-of-10-dimensions/

2 Hugh Ross. The Genesis Question: Scientific Advances and the Accuracy of Genesis. NavPress and Reasons to Believe, 2001. Page 17 - 25.

3 Ibid

4 Ibid, 27 - 34

5 Ibid

6 Ibid

References

Chapter 6:

1   Herodotus, Hist. i.128; iii. 132.2, 159.1

Chapter 7:
1 Fact sheet. "Children: reducing mortality" World Health Organization, updated September 2016
    http://www.who.int/mediacentre/factsheets/fs178/en/
2   Ibid

Chapter 8:

1 T.C. Mitchell, "Geology and the Flood," in New Bible Dictionary, 2nd ed., eds. J.D. Douglas, et al.
(Wheaton, IL: Tyndale, 1982), pages 382-383

2   Ghosh, Pallab. "'First Human' discovered in Ethiopia". BBC News website, Science & Environment.
    4 March 2015 < http://www.bbc.com/news/science-environment-31718336>
3   Krulwich, Robert. "How Human Beings Almost Vanished From Earth In 70,000 B.C."
Krulwich wonders, npr CPR NEWS. October 22, 2012
http://www.npr.org/sections/krulwich/2012/10/22/163397584/how-human-beings-almost-vanished-from-earth-in-70-000-b-c.

4   Krulwich, Robert. "How Human Beings Almost Vanished From Earth In 70,000 B.C. Krulwich wonders,
npr CPR NEWS. October 22, 2012
http://www.npr.org/sections/krulwich/2012/10/22/163397584/how-human-beings-almost-vanished-from-earth-in-70-000-b-c.

5 Zondervan NIV Study Bible, Genesis 11:3 study notes. Grand Rapids, MI: Zondervan Corporation,
1985

6 Zondervan NIV Study Bible, Genesis 11:4 study notes. Grand Rapids, MI: Zondervan Corporation,
1985

7 Ker Than. "On Way to New World, First Americans Made a 10,000-Year Pit Stop". National
Geographic news. February 27, 2014
http://news.nationalgeographic.com/news/2014/04/140227-native-americans-beringia-bering-strait-pit-stop/

8 Hugh Ross. The Genesis Question: Scientific Advances and the Accuracy of Genesis. NavPress and Reasons to Believe, 2001. Page 123 - 125.

9 Ibid, 119 – 123

Move to Reality

INDEX

Index

Index

NOTES:

Index